W0037349

A Philosophical View of the Ocean and Humanity

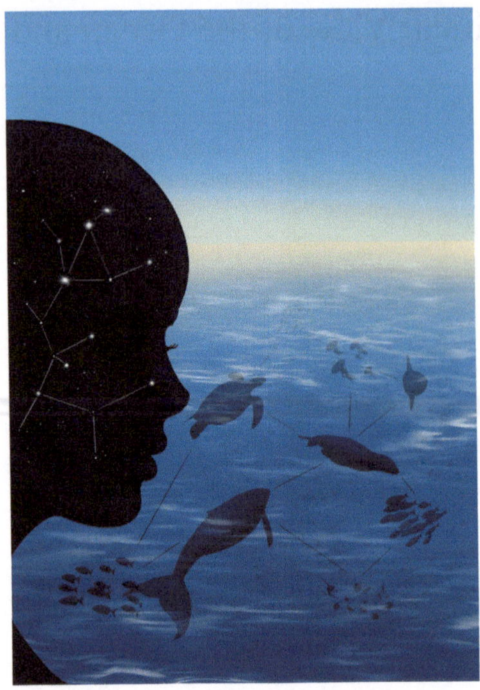

Our inner ocean of rational thinking, feelings, thoughts, and values holds a powerful ability to find a new relationship with the sea, while all the voices of the outer ocean, in the form of sounds from many marine sources, observations, literature, art, and music, call for our attention. *Illustration* Jan Heuschele

Anders Omstedt

A Philosophical View of the Ocean and Humanity

Second Edition

Anders Omstedt
Department of Marine Sciences
University of Gothenburg
Gothenburg, Västra Götalands Län, Sweden

ISBN 978-3-031-64325-5 ISBN 978-3-031-64326-2 (eBook)
https://doi.org/10.1007/978-3-031-64326-2

This Springer imprint is published by the registered company Springer Nature Switzerland AG
The registered company address is: Gewerbestrasse 11, 6330 Cham, Switzerland

Foreword by Bernt Gustavsson (Second Edition)

We know that we live in a time that is problematic in many ways. But we know less about how we should relate to the problems that surround us. In the mass media noise where everything seems to be equally important, science, despite being well developed, has difficulty making itself heard. Slightly simplified, we can say that science is divided between the study of nature and its conditions and man and our society, a division into the external and the internal. This division is as old as the rise of science and philosophy, and it still persists. It lies in the very organization of science—in natural science—and the human and social sciences, and it lies in the very division of philosophy into analytical philosophy and philosophy of life or existence. Plato introduced a division of the world that is usually called dualism, which consists in separating the material from the ideal. When we know, but seemingly nothing happens, a doubt sets in that sometimes borders on despair. We suffer from the absence of action possibilities; we see the real but miss the possible. Therefore, we need tools that make us see different perspectives. Strict science needs to ally itself with art, with literature, with that which increases our possibilities of action so that we know what we do not yet see.

Anders Omstedt's book takes up this challenge by connecting the description of the ocean from the perspectives of both science and art. This productive synthesis encourages us to develop our thoughts about a new relationship to the sea. In the book, the author dives into the outer ocean, with all its many challenges, and into the inner ocean, with access to intuition, art, and dreams. Anders Omstedt illustrates what is perhaps the most important thing in a time of enormous human pressure on the sea: to examine sources of knowledge available in science and art, to dive deep below the surface, to try to understand, to see what is not directly visible, to use knowledge and imagination, and to awaken ingenuity and curiosity. The book shows how the beauty and vulnerability of the ocean and humanity can inspire us to change attitudes that can lead to improved health and harmony for both the ocean and humans.

Trondheim, Norway   Bernt Gustavsson

Bernt Gustavsson Professor Emeritus in Education and Democracy at Örebro University and Pedagogy and Philosophy at the Norwegian University of Science and Technology, Trondheim.

Foreword by Markus Meier (Second Edition)

Alexander von Humboldt (1769–1859) pioneered ecological thinking. For him, the insight was: "Everything is interaction." Before him, many prominent scientists, such as Copernicus, Galileo, and Newton, had increased the understanding of the world during the transition from the Middle Ages to the modern era. Their discoveries had great significance for the development of natural science, for example, the equations of motion in today's climate and ocean models are based on Newton's laws. In the 1960s, research on the earth system began; a holistic perspective for understanding global environmental change emerged through the realization that the earth is a coherent dynamic system that can only be understood if one takes into account the interplay between land, atmosphere, water, ice, biosphere, societies, technology, and economy. Environmental changes can have a natural origin or be influenced by human behavior. The latter include the use of fossil fuels, urbanization, land use, fishing, and waste production.

As an important part of the earth system, the ocean is important to the climate due to its role as heat, water, and carbon reservoirs. It also has great economic importance as a source of food and energy, transport routes, or recreation. For many people, the ocean is also a metaphor for endless expanses, adventure, and pristine and fascinating underwater nature, which is reflected, for example, in the nature films of Jacques-Yves Cousteau.

This is where Anders Omstedt's book comes in, where he describes man's inner and outer oceans and connects the two seas by making them communicate with each other and thus change man's relationship to the ocean. Anders Omstedt knows both oceans well. As an oceanographer, he has intensively studied the processes of the outer ocean and the earth system as a whole, using a holistic approach and scientific methods based on those developed by Humboldt and other great naturalists. He shows the great and partly hidden riches of the sea and the harmful behavior of man.

Anders Omstedt has also investigated his dreams, art, and literature. He has identified using emotions, values, and imagination to change our behavior and become more attentive and careful with nature. In this way, he allows the outer and the inner oceans to communicate with each other. This empathetic communication wants to contribute to solving problems, leading to measures being taken to protect the sea, thus contributing to a future humans strive for.

Rostock, Germany Markus Meier

Markus Meier Head of the Department of Physical Oceanography and Instrumentation at the Leibniz Institute for Baltic Sea Research Warnemünde, and Professor at the University of Rostock, Germany. Chairman of the Baltic Earth Science Steering Group.

Foreword by Alice Newton (First Edition)

Late in the twentieth century, hopefully not too late, humankind began to realize that planet earth was really planet ocean. Slowly, we are coming to understand the importance of the ocean in the earth system. The ocean regulates our climate and has so far buffered some of the worst effects of climate change, aggravated by our greenhouse gas emissions. Nevertheless, our ignorance of the ocean is profound, the gaps in our knowledge unfathomable.

An increasing proportion of the world's population lives on coastlines, narrow, vulnerable strips of land bordering our continents and is subject to storm surges, tsunamis, erosion, subsidence, and sea-level rise. Coastal communities depend on the ecosystem services of the ocean to provide them with food and energy, protect them from storms, regulate temperatures with cooling ocean breezes, support abundant fisheries, and provide them with places for tourism, leisure, water sports, and enjoyment. Great cities are now turning to the ocean for the provision of water through desalination. We are exploring the energy of waves and currents and the rich minerals of the seabed. We are discovering the medicinal uses of more and more marine organisms in our search for cures for cancer and other diseases.

However, our human activities threaten the very ocean that we depend on as we continue to contaminate, pollute, over-extract resources, dam rivers, and change the configuration of the coastline. As we do so, we destroy the biodiversity and ecosystems that we ultimately depend on, especially the ecosystem engineer communities in mangrove forests, saltmarshes, seagrass meadows, and coral reefs.

Anders Omstedt is a marine scientist who understands the importance of the ocean. He speaks to us as only a very knowledgeable scientist can, in a clear voice that makes complex scientific processes understandable. He also speaks to us as a philosopher, one who thinks deeply about the deep oceans. Finally, he allows himself to become a medium, someone who speaks and becomes the voice of the ocean.

This book links science, art, philosophy, and psychology in a profoundly satisfying way. It touches those who love the sea; it tells the scientist that the ocean is more than a set of processes that can be expressed by equations and box models. Most of all, it reminds us that humans depend on the sea and that we should treat the sea with respect.

Faro, Portugal Alice Newton

Alice Newton Professor in the Department of Earth, Environmental, and Marine Sciences (DCTMA), Centre for Marine and Environmental Research (CIMA), Gambelas Campus, University of Algarve, 8005-139 Faro, Portugal.

Foreword by Martin Visbeck (First Edition)

Humankind is linked to the ocean in many ways: the ocean encircles the planet, covers 71% of the earth's surface, and holds unmatched, and sometimes untouched, natural resources—more than we could ever imagine. Here, life began and has been sustained for eons, harboring secrets we have yet to unlock. For generations, the vast and endless ocean, romantic beachscapes, bizarre deep-sea life, and the interplay of wind, waves, and light have fascinated and inspired artists, writers, and filmmakers. The ocean is a source of food and energy and the highway of international commerce, and 40% of humanity lives within 100 km of its shores. It provides sustenance and livelihoods and links our global economy. But the ocean is also a vast yet final warning system displaying the tell-tale signs of our careless lifestyles: it has become highly stressed by our growing interventions and is suffering under the burden of pollution and climate change.

From a scientific perspective, the diagnosis is clear: we need to re-evaluate current human–ocean interactions. Natural science provides mounting evidence of how collective human activity is changing the ocean, putting dramatically increasing pressure on its ecosystem and threatening a number of vital ecosystem services. As well, the social sciences and humanities are beginning to understand the deep emotional and societal connections between humans and the ocean.

A critical challenge is to reconcile these perspectives and develop a scientific systems understanding of possible sustainable development pathways. What societal and technical transformations are needed to safeguard critical marine ecosystem services, and what governance arrangements are needed to share ocean prosperity and benefits globally? How can we be good stewards of our planet so that we and future generations can live in harmony with the ocean? How can we use our understanding of the ocean and human systems to change our actions to support human life within safe limits so that the ocean can continue to sustain its ecosystem services for us as it has for so long?

The recently proclaimed United Nations' Decade of Ocean Science for Sustainable Development (2021–2030) provides a rare opportunity for marine scientists to show how working closely across disciplines and with societal actors to co-design innovative and transformative solutions can lead to a more sustainable human–ocean relationship. Anders Omstedt's book takes on this challenge by describing the ocean from the dual perspectives of science and the arts, a productive synthesis that encourages us to develop our ocean system thinking. It motivates us, providing the hope needed to take transformative human action to safeguard the **ocean we need for the future we want**.

Kiel, Germany Martin Visbeck

Martin Visbeck Professor of Physical Oceanography at the GEOMAR Helmholtz Centre for Ocean Research Kiel and Kiel University, and spokesperson of the Kiel "Future Ocean" cluster.

Preface to the Second Edition

The first edition is rewritten with new illustrations and presented in Part II. The second edition also adds two new parts to the first, Parts I and III. The aim is to introduce the reader to humans' challenges about the ocean and the future. It addresses one of the main questions in the United Nations Ocean Decade that aims to change how humans deal with the ocean. The book stimulates a broad way of thinking by connecting analytical science thinking and intuition. In the book's first part, art and dreaming are used to relate to science. This knowledge is applied in Part II of the book, written in two modes: a concerned science mode and an intuitive, artistic mode in which the ocean is given a voice. Part II illustrates how science and art can be connected to increase awareness of the state of the ocean and support behavioral change.

Part III deepens the description of humans' relationship to the ocean and our way of thinking with inspiration from literature and philosophy. Various possibilities for changing behavior are discussed there. By paying attention to the voices of the ocean embodied in art and collected through extensive observations by marine scientists, there are opportunities for a new relationship. There are unlimited amounts of hidden intelligence to discover and be inspired by. All the voices of the ocean today call for our attention.

Gothenburg, Sweden　　Anders Omstedt

Preface to the Second Edition

Preface to the First Edition

By the end of this century, it is estimated that the human population will have grown to over 10 billion people, and by 2050, almost 70% are expected to be living in urban areas and megacities increasingly alienated from the marine environment with its pervasive plastic contamination. Today's young generation will need to respond to many of today's alarming warning signs. Perhaps the most important question to be addressed is how humans can direct marine development onto a sustainable path while preserving their humanity.

Early in my study of the oceans, I became interested in the thermodynamics of the water surface layer and ice formation. The initial ice formation in the ocean in the form of Frazil Ice was the topic of my Ph.D. research. At that time, I became increasingly aware that my studies of the external ocean also fed into my studies of my "internal ocean," and I was relieved to find that my emotional awareness was not frozen. Frazil Ice is much more dynamic than solid ice and served as a metaphor for my emotional life—wild and fascinating.

Later, during post-doctoral studies in Canada, I studied deep water processes in Lake Ontario and the Baltic Sea while becoming interested in my own and others' dreams. I became aware that I could think in different ways. Working with computer coding and answering children's questions on the same day gave me experience of how the left and right sides of my brain worked and complemented each other—though this was not discussed among natural scientists then. I was becoming interested in how others and I thought about and processed experiences.

My oceanography work became increasingly oriented toward understanding systems and modeling physical ocean processes fed into biochemical modeling of the carbon system. This opened up the possibility of addressing problems related to climate change and eutrophication and modeling multiple ocean stressors. Such modeling, mostly using a bottom-up approach, became more and more complex. The human effects on the climate were obvious. They could be modeled by considering past and present climatic conditions, extending them into the future by prescribing different emission pathways. Anthropogenic pressures on the ocean, especially its coastal seas, are strong in various ways, opening up new questions about how to model human impacts.

What determines human behavior and what pathway humanity will take in the future are questions that cannot be answered with certainty. It was obvious to me that science needs to improve its understanding of human behavior and perceptions of the ocean, so I wrote a book about how analytical thinking and intuition could be better and more productively connected. Dreams can be used as a teaching tool in transforming emotions into stories of great value and psychological resonance, and these stories can be used to integrate analytical thought and intuition. From long experience in dream group work using the Ullman method, I realized that dream analysis provides an excellent background for studying our thinking. I argued that scientists should connect science and the arts better. In this book, I expand on this idea by investigating the connection between science and the arts, starting from marine science and our conscious and unconscious perceptions of the ocean. The aim is to illustrate the central importance of the ocean for humans and expose the disconnect between the ocean and human emotions that leads to the misuse of the former.

Gothenburg, Sweden Anders Omstedt

Acknowledgments

Over the years, many colleagues and students have inspired me to deepen my knowledge of the ocean. Many talented dream group leaders and participants have passed on their knowledge of dream work and literature. The book supports Gothenburg University's interdisciplinary marine research, Baltic Earth, and the UN's Ocean Decade to stimulate reflections on how our relationship with the ocean can change. Scientifically, I have greatly enjoyed my oceanography and meteorology work and have met many independent and creative researchers. Work in the BALTEX/Baltic Earth programs has been my main school for international work, and there, friendly ties have been built around the entire Baltic Sea with researchers from different parts of the world. The BALTEX/Baltic Earth International Secretariat at Helmholtz-Zentrum Hereon has strongly supported the organization. I would also like to thank Stina Hammar for introducing me to a new way of working with literature and dreams and members of the Educational Forum and Dream Group Forum for many fruitful meetings.

Introduction

The ocean is, for many, a rich source of joy and inspiration. It forms a unit with connections to all the coasts of the earth. At the same time, we subject both the ocean and its coastal seas to enormous stress. From being considered an infinite resource, free to explore and exploit, we now know that the sea is threatened and lacks adequate protection. Extensive international initiatives are underway, such as the UN Decade of the Oceans (2021–2030), where the ambition is to take measures to "safeguard the ocean we need for the future we want," and Agenda 2030, where the goal is to create conditions for global Sustainable Development. Here, we face great challenges linked to technical/natural science knowledge of the ocean and the human sciences' knowledge of our needs and behavior. Mere facts about the state of the oceans are not enough. Feelings and values are also needed here, something that has its source in our inner oceans. In the book, the ocean and humans' hidden intelligence are raised as inspiration for change. With its great and hidden riches, the world ocean is represented as the outer ocean, and the human psyche, with its inner universe of feelings, values, and imagination, as the internal ocean. The communication between the outer and inner oceans occurs through many different voices, where science, art, and the extensive measurements of marine scientists call for attention.

The book's Part I introduced the reader to how analytical thinking and intuition can be connected. Here, we use poetry and dreams to examine how fleeting feelings can be translated into stories of great value. Our imagination and our dreams form the basis of all human creation. If society has an overconfidence in rationality, we are deprived of the possibility of developing these inner resources. The book's Part I aims to increase the understanding of intuition and how we can rationalize and intuitively think.

This knowledge is used in Part II of the book, where marine science is connected to the arts through communication between a marine scientist and the ocean. In the dialog, we are reminded that the most meaningful part of life is nurturing new life under healthy living conditions. Developing healthy and joyful living conditions is possible, and with the growing human population, it should be humanity's primary task. Destructive thinking and behavior scare us and will not promote the future development of a more sustainable lifestyle. Human growth and culture brought about by art, literature, and science have extraordinary capacities and can provide the impetus for necessary mental change, as illustrated by humanity's long history of daring achievement. Nothing, except human nature, has been too difficult to overcome, but the price has often been high.

In Part III of the book, the discussion about the need to connect science and art and create the conditions for many voices to be heard is broadened. The development of the ocean and the earth has been going on for billions of years and formed a blue planet with unique conditions for life. Only recently have humans come to influence nature in many negative ways. However, today, there is a completely different knowledge and conditions than previous generations, which means we can better manage the oceans. The ocean cannot take care of itself, and we cannot restore it to how it was under previous conditions. We must realize that the ocean's future is closely linked to what humans do. By paying attention to the voices of the ocean embodied in art and collected through extensive observations by marine scientists, there are opportunities for a new relationship. With access to the scientific source of knowledge and inspiration from the artistic source of knowledge, there is hope and insight that a fundamentally new way of thinking is possible, where humans and the ocean can be united in a new relation that can guarantee vitality and progress.

Contents

About the Author

Anders Omstedt is a professor emeritus in oceanography at the Department of Marine Sciences, University of Gothenburg. In 2001, he took up the Swedish Research Council's professorship in geosphere dynamics. Omstedt has extensive experience in marine sciences, especially in modeling related to processes regarding:

- sea ice dynamics and thermodynamics
- water temperature
- salinity
- turbulent mixing
- energy and water balance
- marine biogeochemical processes
- oxygen dynamics
- nutrient dynamics
- carbon dynamics.

His work has included the development of ocean and sea ice forecast models, model studies of climate and climate change, ocean acidification, and eutrophication. Omstedt was a science coordinator at the Marine Environment Institute and held various positions at the Swedish Meteorological and Hydrological Institute (SMHI). He has been chairman of the Baltic Sea Experiment (BALTEX) and shared the chairmanship with Hans von Storch for the BALTEX Assessment of Climate Change for the Baltic Sea Basin (BACC). In parallel with his research, Omstedt has trained in literature, psychology, and dreams by participating in/or leading many courses around Sweden. He is a dream group leader in the Swedish Dream Group Forum activities.

About the Author

List of Figures

Part I
And the Nights Abound with Inspiration

Chapter 1
Opening, Part I

Abstract This opening chapter illustrates how complex problems of reality often need to be addressed from many different perspectives, where science, arts, and intuition all belong to sources of inspiration and tools for problem-solving. How analytical and intuitive thinking can be connected is addressed by observing how we think and feel. The ocean is a source for analysis of the human condition in all its complexity. The need to think again and balance scientific and humanistic values increases today with the many complex problems that society needs to solve. We will use marine science and poetry, dreams, and imagery tools for inspiration here. The ocean is a strong symbol of our inner unconscious. The sea, as the arts, can awaken inspiration, feelings, and memories, remind us of what is important in our lives, and open the possibilities for behavior change.

Most people know that when struggling with a problem, the answer may come more easily after a good night's sleep. While sleeping, our minds organize the impressions gathered in the daytime. It is easy to find examples of brilliant scientists who made their discoveries inspired by dreams. For example, after hard work and a dream, Dmitri Mendeleev (1834–1907) organized the periodic table of chemical elements based on their atomic numbers. In daily life as well, a good night's sleep can help us solve many problems. Many of us have learned that creativity is associated with hard work. This is also what we teach our students when loading them with more and more work during the education process. It has surprised me that we scientists and teachers rarely discuss the creative process and how to explore it. Even today, this matter is often not even included in university curricula. Most people are interested in the creative process and have varied idiosyncratic views of how it can be improved, but there is little systematic discussion of how to develop methods to promote creativity. Instead, we feel driven to explore, in an unplanned, unsystematic, un-self-aware way, various inspiration initiatives or to apply suggested new pedagogical methods without deepening our knowledge of creativity.

One purpose of this, the book's first part, is to illustrate how analytical thinking and intuition can be trained by observing how we think and feel. Here, we will use poetry, dreams, and imagery as an intuitive or artistic thinking tool. Another purpose

A. Omstedt, *A Philosophical View of the Ocean and Humanity*,
https://doi.org/10.1007/978-3-031-64326-2_1

is to show the beauty and power of how our subconscious and creative thinking support our conscious self. The presentation will follow a method developed by Montague Ullman (1916–2008), an American professor of psychiatry (Ullman, 1996; Siivola, 2011). In Ullman's experience-based dream group method, which has also been applied in Sweden for many years, analytical/rational thinking is successfully combined with an intuitive/artistic way to reflect.

Part I of this book explores poetry and dreams. They have been taken from literature and my own experience (poetry and retold dreams are written throughout the text in italics). These are intertwined with my scientific experience as a marine scientist (Omstedt, 2016). Scientific work can be seen as exploring the unknown using our conscious senses. Systematic work in the marine sciences requires many years of training and a willingness to ask new questions. Working with dreams can be seen as artistic work, done with a language other than analytical thinking, and often expresses emotions strikingly and emphatically. It is a resource freely available for all to explore and a gateway to a deeper understanding of ourselves, humanity, and creativity.

Our various sources of knowledge can be divided into:

- The scientific,
- the sensual or artistic,
- and the mythological.

How these are used in our thinking can be illustrated by starting with the importance of the ocean as a natural resource and source of inspiration. The sea and its ecosystem are intertwined with humans through an extensive network of services and emotions and contain many conditions for life. The importance of the ocean for humans is made clear, not least by the many poems and stories that deal with the sea. For example, in the first song of the Finnish poem Kalevala, Ilmatar, the maiden of the ocean and the air, gives birth to a son, Väinämöinen:

……..

Väinämöinen was born alone,

Greatest among singers, to the world;

Ilmatar was the maiden's name,

Who became the mother of Väinämöinen

….

The wind rocked the maiden.

Swept away by the waves,

Drove far out on the waves of the bay,

Dragged by white foaming waves;

The wind did what it did;

she became pregnant with the sea.

(Elias Lönnroot, Kalevala 1999, translated into English).

How can this text be understood? If we choose a scientific perspective, we are blocked by questions about how the sea can become a father. If we overcome this, we may be inspired to think about how life was created in the ocean and how sunlight-driven photosynthesis in phytoplankton converts carbon dioxide and water into life and oxygenates our atmosphere. With an artistic perspective, we can approach this text more freely and content ourselves with letting it depict the beauty of how the atmosphere and ocean come together. If we allow ourselves to go to the mythological source of knowledge, we can rejoice in what this story teaches us: that in a change of perspective—Ilmatar leaves the sky and descends to the ocean—something new can be born.

The ocean is a strong symbol of our inner unconscious. Deep and vast, it holds the conditions of life, and on the surface, there is the consciousness (Fig. 1.1). The ocean can thus awaken inspiration, feelings, and memories in us and remind us of what is important in our lives. Perhaps the ocean can also, so exposed to uncontrolled destructiveness, symbolize the collective psyche of humanity and how we do not get the mental nourishment we need.

Döring and Winther (2021) introduced the analogy: *The human condition is an ocean.* The ocean is a source for analysis of the human condition in all its complexity—emotion, freedom, sexuality, imagination, dream, memory, political structures, and cultural conditioning. Analogy means comparing an object in one domain with another and opening us to new perspectives. In this case, the authors state that we can learn much about human conditions from the ocean.

Scientists today strive to develop more and more complex models for how humanity affects nature. But, no models can describe the ecosystems, the role of humans, and what may happen in the future. The need to think again and balance scientific and humanistic values increases today with the many complex problems that society needs to solve. Much work has been concentrated on external phenomena, but the importance of the human's inner world has been neglected (e.g., Ives et al., 2020). For the development of a more sustainable relationship between people and nature, the inner life must, therefore, be considered as a player in a changed outer life, which is also the theme of this book.

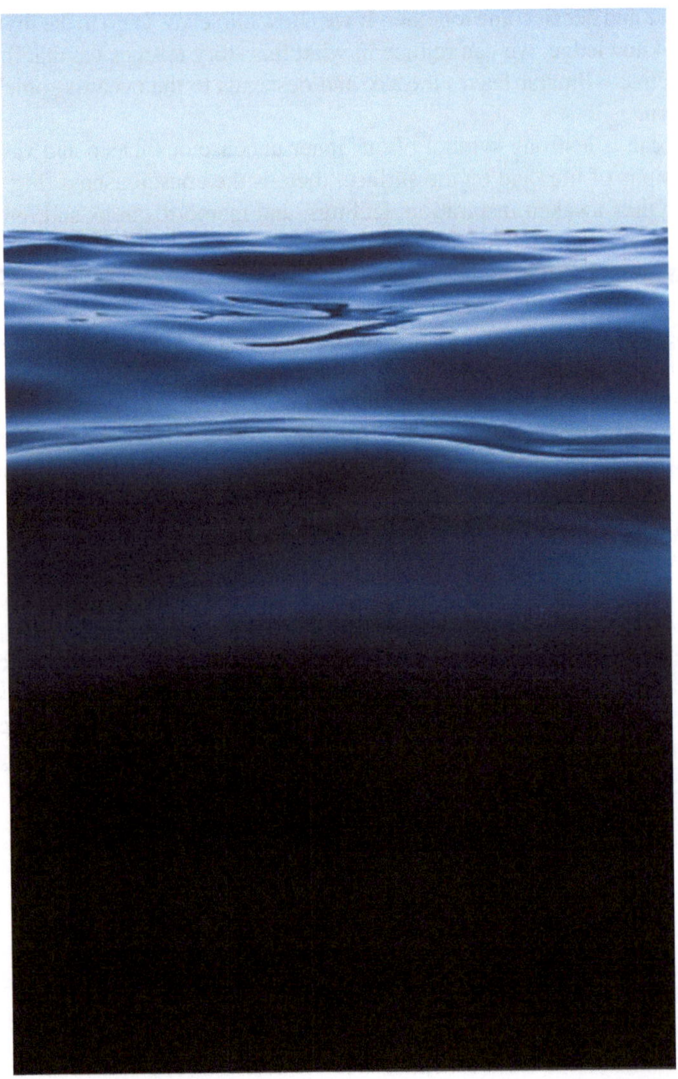

Fig. 1.1 We detect only movements on the surface when we look out over the ocean. Deep in the sea are various movements, hidden resources, and sound waves that travel effectively. *Photo* Hillevi Nagel

References

Döring A, Grønfeldt Winther R (2021) The human condition is an ocean: Philosophy and the Mediterranean Sea. Chapter in: words and worlds: use and abuse of analogies and metaphors within sciences and humanities. In: Wuppuluri S, Grayling AC (eds) Synthese library series. Springer

Kalevala (1999) Finland's national epic. Finnish folk poems compiled by Elias Lönnrot. Translation by Huldén L. Huldén M (in Swedish). Atlantis, Stockholm

Ives CD, Freeth R Fischer J (2020) Inside-out sustainability: The neglect of inner words. Ambio 49:208–217

Omstedt A (2016) Connecting analytical thinking and intuition: And the nights abound with inspiration Springer briefs in Earth sciences. Springer, Cham, Switzerland

Siivola M (2011) Understanding dreams. The gateway to dreams without dream interpretation. Cosimo, Inc., New York, USA

Ullman M (1996) Appreciating dreams: a group approach. Sage Publications Inc. International Educational and Professional Publisher. Thousand Oaks, California, USA

Chapter 2
Searching for a New Language

Abstract The language of the arts can provide a different way to address the reality. Rational thinking using science is a strong instrument for developing a problem-oriented language but often misses all feeling colors. Here, the start is to search for the art language to help balance the rational language. Dreams and art can help open up this by asking questions such as: What are the feelings that are awakened, and what is the meaning of the dream or art images?

When I was around thirty, I needed to develop a new language to express my feelings and ideas better. I felt frozen and trapped by many expectations. The Swedish poet Dagerman (1954) helped me put words to my feelings in his poem *Ice Age*:

You have no soul when it's cold,
You have an icy little business stamp.
You might almost wish you could melt.
And become a river, like the Nile, for example.

But here, no one becomes what he wants to be.
You stand like a statue in the snowy landscape.
Suddenly, one day, you have frozen solid
and every time you open your mouth, you spout ice.

And then everything happens as foretold.
In front of some fire, your icy profile melts.
The hostess brings nothing but a rag.
You become a puddle, though you wanted to be the Nile.

The poem was written before the Aswan Dam was built to regulate the Nile River, the world's longest river. Before the dam was built, the Nile River flooded every year, bringing water, nutrients, minerals, and silt to fertilize and build up the delta, making the Nile Valley and Delta ideal for farming. Today, a combination of regional anthropogenic changes, such as the Aswan Dam and climate change induced by increased greenhouse gas levels, is seriously threatening the Egyptian coast. Later, I worked on climate change in the Mediterranean Sea with Mohamed

A. Omstedt, *A Philosophical View of the Ocean and Humanity*,
https://doi.org/10.1007/978-3-031-64326-2_2

Shaltout from the University of Alexandria. We demonstrated that even a slight rise in sea level could greatly damage the Nile Delta region (Shaltout and Omstedt 2015), resulting in a need to organize effective integrated coastal zone management to prevent future flooding. Society outside science needs to be involved in devising this management. Such interaction is often required in the Earth Sciences, creating a need for new communication skills not touched on in most science curricula but discussed widely within climate change. In my early thirties, I wrote a short article in our union newsletter, accompanied by Stig Dagerman's *Ice Age*, expressing my frustration of being trapped in an old working organization. My search for a new language had started.

I defended my Ph.D. thesis about cooling and ice formation in the ocean while working at the Swedish Meteorological and Hydrological Institute (SMHI), developing forecasting models for the ice-breaking service. The scientific work was a great pleasure; in particular, testing theories about the ocean against field observations evoked strange and exciting feelings (Fig. 2.1).

The possibility of integrating mathematical expression into computer codes and testing theories directly against observations inspired me. However, I did not feel satisfied or free and needed to redirect my life. I attended a summer course to work on literature, dreams, and myths. The first night, I dreamt of standing outside a department store and seeing *lots of interesting boxes inside. When I went to the main entrance, the door was locked.* The night after this dream, I dreamt of being *bitten by a scorpion on my Achilles tendon.* My interest in dreams awakened, and I realized that my vulnerability and feelings were the key to my dreams. That first summer course was at Biskops-Arnö, a beautiful place on Lake Mälaren outside Stockholm, organized by Stina Hammar (1923–2020), who has written several books about literature and dreams. From that time, attending summer courses at different places around Sweden and working in dream groups gave me an education in literature and dream interpretation. At the same time, I was exploring my academic field and deepening my understanding of how the ocean and its coastal seas function.

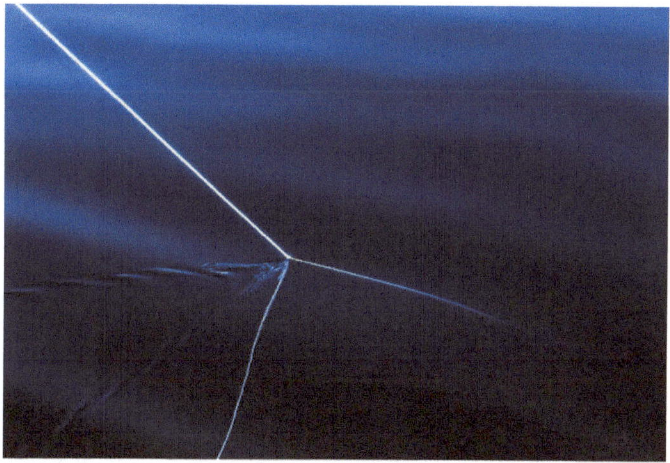

Fig. 2.1 Line for measurements in the Gullmarsfjord on the Swedish west coast. *Photo* Christian Stranne

References

Dagerman S (1954) Newspapers: published in the newspaper Arbetaren 1944–1954. Nordstedt förlag published these in 1983, where the poem Istid is included (in Swedish)

Shaltout, M, och Omstedt A (2015) Modelling the water and heat balances of the Mediterranean Sea using a two-basin model and available meteorological, hydrological and ocean data. Oceanologia 57:116–131

Chapter 3
Finding Words for Feelings

Abstract Here, we start to examine the language of a dream from an artistic perspective, where emotions, symbols, and metaphors are the bases for understanding. Intuition, dreams, and fantasy are internal sources that can give a new way of thinking. Finding words for feeling is healing and opens up new ideas. Connecting rational thinking and intuition can create a strong instrument to process reality and inspire behavior change.

> *I am out sailing in a large modern sailing boat. We have strong winds behind us, and all the sails are unfurled. I am on deck tending to the sails. Suddenly, I see my young son climbing onto the deck and falling into the sea.*
>
> *I rush to the tiller and turn around the sailing boat. A man helps me bring my son on board.*

Dreams make it possible to plumb deep into one's unconscious inner ocean and capture images that transfer information from our hidden resources to the surface. Putting words to one's feelings is important because feelings are involved in communication between the deeper and more superficial parts of our psyche. Finding words for feeling can initiate healing and unlock creativity. This is well known in psychology and has been described by, e.g., Marie Cardinal (1929–2001) in her famous book from 1975, translated into Swedish in 1978, Cardinal, 1978).

The dream above evokes feelings of happiness to sail at full speed, wildness, stress, worry, fear, agency, and liberation. At the same time, the dream is full of symbols and metaphors with many different meanings. Sailing at full sail is a metaphor indicating that the dreamer's full capacity is being used and life is moving fast. A son falling into the sea is another metaphor that may tell the dreamer is in danger of missing something important. Saving the son with the help of another man, another resource in the dreamer, suggests that the dreamer should pay attention to another aspect of himself. When I had this dream, I had just started a project with a friend and a small group of people. The first meeting had been a disaster, and the group seemed heading in two different pedagogical directions. After this dream, I realized I had to talk to my friend and take over the project management. This project was important to me,

A. Omstedt, *A Philosophical View of the Ocean and Humanity*,
https://doi.org/10.1007/978-3-031-64326-2_3

and the group could not go in two directions (Fig. 3.1). My friend admitted she was tired and happy that I wanted to lead the project. In this dream, my feelings guided me to understand the situation better and improve my leadership.

Fig. 3.1 How can we navigate in a complex world? *Photo* Hillevi Nagel

Reference

Cardinal M (1978). Orden som befriar (in Swedish). Trevi, Stockholm. ISBN 9171603468. The original title is Les Mots Pour le Dire from 1975. It is available in English and is titled The Words to Say It, from 2013

Chapter 4
What Are Dreams?

Abstract Dreams have been studied along with human history and given different interpretations. Some of the ideas are reviewed in this section. They should not be treated as if they are manuals for wisdom. Instead, dreams can be seen as a kind of internal communication that, through feelings and images, transfers information between our unconscious and conscious selves. The dream images are full of symbols and metaphors, i.e., a representation of something else, and can be given many interpretations.

Much has been written about dreams from many different perspectives. In particular, Sigmund Freud (1856–1939) and Carl G. Jung (1875–1961) have strongly influenced our understanding of dreams. The psychologist Eric Fromm (1900–1980) describes dreams as a forgotten language (Fromm 1982). A language that everyone has access to and is therefore universal. Anthony Stevens (1933–), a doctor and psychoanalyst, reproduces dreams from older and more recent times and analyzes some famous dreams, for example, by Hitler and Descartes (Stevens 1997). John Sanford (1929–2005), a priest and Jungian analyst, calls the dreams that appear in the stories of the Bible the forgotten language of God (Sanford 1989). Ole Vedfelt (1941–), an analytical psychologist, provides a comprehensive modern overview of dreams and dreamwork (Vedfelt 1996). Dreams can be seen as a kind of internal communication that, through feelings and images, transfers knowledge between our unconscious and our conscious self. The dream images are full of symbols and metaphors, i.e., a representation of something else, and can be given many interpretations.

Interestingly, the opposite of a symbol in the mythological source of thought is diabolo, where a symbol holds together, and diabolo is something that divides. Two of the climate models' calculations for the earth's future climate development are a green and a fragmented world, respectively. It is easy to imagine that the first stands as a symbol of a more human world and the second for a destructive world. However, the symbols and metaphors of dreams do not have a standard explanation but can mainly be understood by the dreamer himself, preferably with the help of others. Below is an example of another dream:

A. Omstedt, *A Philosophical View of the Ocean and Humanity*,
https://doi.org/10.1007/978-3-031-64326-2_4

I am steering a motor boat from the Baltic Sea into the Kattegat.
Several of us are on deck. Suddenly, we see a military ship coming towards us.
It comes close to us, and a Nazi officer jumps onboard.
Our boat has been seized.
Inside the cabin, the crew starts to organize resistance.
The children start an orchestra using pots and kitchenware.
and sing along with their playing.
The Nazi officer leaves the boat in fear, and we are free.
I resume steering the boat and carefully watch the sea.

Dream symbols can illuminate feelings in dramatic and comic ways (Fig. 4.1). At about the turn of the century, I had the above dream just after I moved from one job in Norrköping, a city near the Baltic Sea coast, to another job in Göteborg, a city on the Kattegat coast. The motorboat symbolizes my situation in life, traveling from one city to another. The symbol of the Nazi military ship occupying my life illustrates the pressure I was experiencing in taking this new job. The children's actions characterize how I defended my personality, symbolizing a different and more creative approach to life. The dream bolstered my confidence in my new work and told me to be careful and watch out for occupying and difficult feelings. The dream illustrated two attitudes inside my mind, controlled and playful, and the importance of understanding the difference between them.

Many dreams are not easy to understand for the individual dreamer. It is a common mistake to quickly try to interpret one's dreams, a mistake because it often leads to a negative interpretation of the dream. With the help of dream group work instead, a rich opportunity opens up to better understand and anticipate the language of dreams. The dream often shows a deep truth in the dreamer and supports mental health. In this way, new ways of thinking can be created that enable problem-solving and reorientation, i.e., improved thinking where emotions and reason are brought together innovatively.

Fig. 4.1 What emotions does the ocean evoke? *Photo* Hillevi Nagel

References

Fromm E (1982) The forgotten language (in Swedish). Nature & Culture, Stockholm.
Sanford JA (1989) Dreams. God's forgotten language. Harper Collins Publishers, New York
Stevens A (1997) Private myths. About dreams and dreaming (in Swedish). ScanBook AB Falun
Vedfelt O (1996) Dreams and dream interpretation, Swedish edn. Forum

Chapter 5
Meaning of Symbols and Metaphors

Abstract Symbols and metaphors are everywhere: in science, art, religion, literature, and dreams. They act as messengers of otherwise elusive messages and can open new avenues of thought. The word symbol itself comes from the Greek and means sign. Metaphors, which means transfer, originate also from Greek. Stories and metaphors can have something in common. The function of the story is to bridge the gap between the ordinary and the extraordinary or unexpected, unpredictable. In the same way, metaphors build a bridge between the concrete, known, and the abstract, unknown.

Symbols and metaphors are everywhere: in science, art, religion, literature, and dreams. They act as messengers of otherwise elusive messages and can open new avenues of thought. The word symbol itself comes from the Greek and means sign. Simple symbols can be, for example, the Greek letter π, in mathematics called Archimedes' constant, which represents the ratio between the circumference and diameter of a circle. Its value is approximately 3.14159. The number is irrational and cannot be fully written down but elegantly summarized by the symbol π. Other simple symbols are our traffic signs, where the number on the sign tells how fast one can drive. Abbreviations such as UN, EU, and UNESCO can also be seen as symbols. A slightly more complex symbol is, for example, a picture of a windmill, which can have the same meaning as wind and air (Fig. 5.1). The word icon, which in Greek means image, has been used for holy images within the Orthodox Church for millennia. Nowadays, the word is used in many contexts, and in the computer world, it is used as desktop images representing an underlying computer program. Carl Gustav Jung (1875–1961) and collaborators published 1964 a book entitled *Man and His Symbols*, with photos of many interesting symbols (Jung 1964). Symbols or imagery help us interpret beyond our reason's limits and can give the sense of a larger context.

A closely related concept to symbols is metaphors, which Gudrun Olsson (1947–), professor of psychology, investigated in her book *In the Landscape of Metaphors—About Life, Death, and Love* (Olsson 2020). Even the word metaphor, which means transfer, originates in Greek. Stories and metaphors can have something in common.

A. Omstedt, *A Philosophical View of the Ocean and Humanity*,
https://doi.org/10.1007/978-3-031-64326-2_5

Fig. 5.1 During the nineteenth century, over 2000 windmills on the island of Öland effectively utilized wind power for grinding grain. In Sweden, they have become a tourist symbol for Öland. *Photo* Anders Omstedt

The function of the story is to bridge the gap between the ordinary and the extraordinary or unexpected, unpredictable. In the same way, metaphors build a bridge between the concrete, known, and the abstract, unknown. In a metaphor, one often distinguishes between the ordinary and the symbolic. For example, in the metaphor "*the day is a desert walk,*" the day is normal, and the desert walk is part of the image. Olsson quotes Aristotle (384–322 BC), who has said that "*metaphor is a word that denotes one thing and is transferred to another thing,*" and Paul Ricoeur (1913–2005) has formulated it like this: "*The metaphor is a spark that brings language to life.*"

References

Jung CG (1964) Man and his symbols. Aldus Books, London

Olsson G (2020). In the landscape of metaphors. About life, death and love (in Swedish). Carlsson Publishing House, Stockholm

Chapter 6
Triggers

Abstract Triggers generate reactions or feelings that influence our psyche and behavior. Certain feelings experienced during the day remain with us. These lingering feelings play an important role in our mental health and often obtain energy from important, unresolved personal issues. We often need to suppress feelings of, for example, stress and anger to be able to cope with daily life. Triggers interact with suppressed feelings and generate dream images that are not about the trigger but about our emotions.

Dreams start from feelings generated in our daily lives but do not end there. Certain feelings experienced during the day remain with us and enter the sleep domain. These lingering feelings play an important role in dream creation and often obtain energy from important, unresolved personal issues. All of us bear the burden of earlier life experiences that make us vulnerable, at the same time as we are facing a future replete with challenges (Fig. 6.1). We often need to suppress feelings of, for example, stress and anger to be able to cope with daily life. This internal stress often triggers dreams, becoming a language of emotions.

My first contact with systematic dream work was at the summer course at Biskops-Arnö, as mentioned in Chap. 2. At that time, I did not know much about dreams, and I was not prepared for dream work when I began the course. The summer school focused on literature, dreams, and myths, and I was only ready for the literature part. On the first day, I met a group of new people and, with an experienced dream group leader, started to work on dreams. A woman presented her dream to the group, and we worked through it during the session. The work was very creative, and the dreamer was open to the process and its potential to alter her understanding of her situation in life. I was greatly impressed. This feeling stayed with me, and in the evening, when I went to bed, I dreamt about *a department store that contained many interesting boxes but had a locked front door*. My experience that day of lacking knowledge of dreams but being impressed by their strength was illustrated by the locked store department containing many potential gifts. In addition, my feelings and experiences before the dream may have triggered the stress of being an outsider to the group. The feeling went into the dream and motivated me to explore my understanding of

A. Omstedt, *A Philosophical View of the Ocean and Humanity*,
https://doi.org/10.1007/978-3-031-64326-2_6

Fig. 6.1 Road of life is full of cracks. Meeting with our subconscious can give us a better perception of reality. *Photo* Hillevi Nagel

dreams deeply. Feelings that were not that different from what one can experience in science when mainstream developments attract the most interest and one decides to go independently.

In my scientific field, oceanography, understanding often starts with simplification, and scientists who have successfully applied this method, such as Henry Stommel (1929–1992), were our heroes. Stommel explained important aspects of Gulf Stream dynamics using simplified mathematical methods. On the other hand, many other scientists, particularly from the marine biology community, glorified concrete details. The book *Siddhartha* by Herman Hesse (1877–1962) tells how a young man deepened his understanding by leaving behind a privileged life to discover his ideas (Hesse 1922). Siddhartha wanted to find the roots of his feelings, believing he could understand the causes of phenomena with this knowledge. He recognized early on that what he knew least about was himself. He embarked on a long journey to learn himself better from that insight. Afterward, Siddhartha formulated the maxim *that for every true idea, there is an opposite one that is also true.* Maybe this also applies to both science and art interpretation. For every good simplification, there is an opposite complex solution that is also true—an exciting possibility that triggered my curiosity.

Reference

Hesse H (1922) Siddharta. Eine indische Dichtung. S. Fischer Verlag. Available in Swedish in libraries

Chapter 7
To Be Touched

Abstract To be touched can generate feelings of responsibility, sympathy, gratitude, or life spirit. Scientists and artists become touched by their problems, which gives energy for many years of work. Poetry, art, literature, dreams, music, and science engage different parts of the human creative faculty and can touch us and give new energy for long-time engagements. To be touched is the key to connecting to life and society, beautifully illustrated by Michelangelo in his fresco *Creation of Adam.*

The strength of being touched is visualized in the famous *Creation of Adam* by the Italian sculptor Michelangelo (1475–1564), illustrating how God reached out and touched the finger of Adam to give him the gift of life. Poetry, art, literature, dreams, music, and science all engage different parts of the human creative faculty, and various aspects of these endeavors can touch us. Perhaps one can say that the joy of learning is one of the most important drivers of progress in life. At the same time, our minds change and sometimes open us up to human kindness. I was first touched by dreams when I dreamt that a scorpion bit my Achilles tendon. Since then, I have studied my and others' dreams for three decades. Parallel to this, I have worked as an oceanographer, exploring how the ocean functions. The strong connections between my scientific work and my dream work have surprised me. Starting with a Ph.D. thesis on cooling and ice formation in the sea, I was struck by a desire to understand how ice interacts with subpolar and polar seas. When modeling ice formation in turbulent water due to the formation of frazil ice (i.e., small ice crystals growing in seawater, Svensson and Omstedt 1998), how ice crystals grow and connect, forming surface ice, first engaged my thinking. Then, understanding how ice drifts, forms ridges, leads, and melts and how this knowledge could be transferred to others using forecasting methods kept me busy for a decade.

At the same time, I started writing down my dreams and studying other people's dreams in literature and dream groups. The theme of several summer courses was to meet oneself through others, and the methods were Ullman's dream group method, literature study, and dance. Intuitive and scientific work gave me strategies to explore the inner and outer oceans. Inspired by a Canadian marine science colleague, I formulated my principles to increase my knowledge, which I later taught students.

A. Omstedt, *A Philosophical View of the Ocean and Humanity*,
https://doi.org/10.1007/978-3-031-64326-2_7

To succeed in science, one must prioritize three aspects. They can be summarized as follows:

- Working on,
- focusing,
- and do not harm yourself.

These three principles gave me a strong foundation for working in science.

A similarly powerful method in dream work is when the group makes a dream of their own and begins to explore the subject's feelings as conveyed in the dream and its symbols. By relating the problem or dream to my own experience and examining and understanding emotions, I learned to systematically improve my understanding of the interplay between facts and emotions. The dream interpretation method increases the capacity for empathy by listening to someone else and relating their experiences to your own. From a philosophical view, this work opens up Hermeneutics, which studies artistic and literary interpretations.

This ability is becoming increasingly important at universities, where researchers need to be able to convey their research results to other experts with different backgrounds, as well as to the public.

Understanding how humans affect nature is an increasingly pressing challenge for the scientific world. Scientifically, I have worked with many researchers and students to understand how the ocean and the climate system work. My scientific approach was based on process understanding and on developing numerical models that successively add processes of great importance, one by one, to build system understanding. By understanding important processes, one could create an understanding of the environment. As researchers, we learned that much was unknown, and we were disappointed that some tried to sell an overly simplistic view. Oceanographers, meteorologists, hydrologists, and geologists first investigated the natural physical system and its connection to chemistry, biology, and human interaction. Our modeling became increasingly complex, and it took my research group over a decade to produce a scientific instrument that could systematically examine human impact on the ocean. From this work, it was clear that humans negatively influence the environment. The message from our scientific work together with many other researchers could be summarized as follows:

Our mentality seems to change when we look at the oceans. On land, there is sometimes a caring mentality, but when we pass the shoreline, we adopt a pirate mentality.

Hidden in our minds are the old assumptions that the oceans are endless and that fish and marine resources are free to explore. The awareness of what we do when we dump waste into the ocean or release it into the atmosphere is blown away (Fig. 7.1). Most people are now moving to cities, and our relationship with the ocean is weakening. We increasingly exploit marine natural resources, forgetting that the ocean represents great resources for our overall development, not only physical resources for food, shipping, and recreation but also imaginative resources that inspire, instill respect, and offer rest (World Ocean Review 1, 2010). The marine management still fails to limit exploitation and cannot preserve the oceans' health. Instead, we have seriously threatened the marine environment in many ways. We must change our behavior and become more attentive and careful with our natural resources. We have entered a new era where nature is very dependent on what we humans do; nature can no longer recover without our support and must get the right conditions for good recovery. Scientifically, we now have methods to guide us toward better management; the limitations are now in society and our minds.

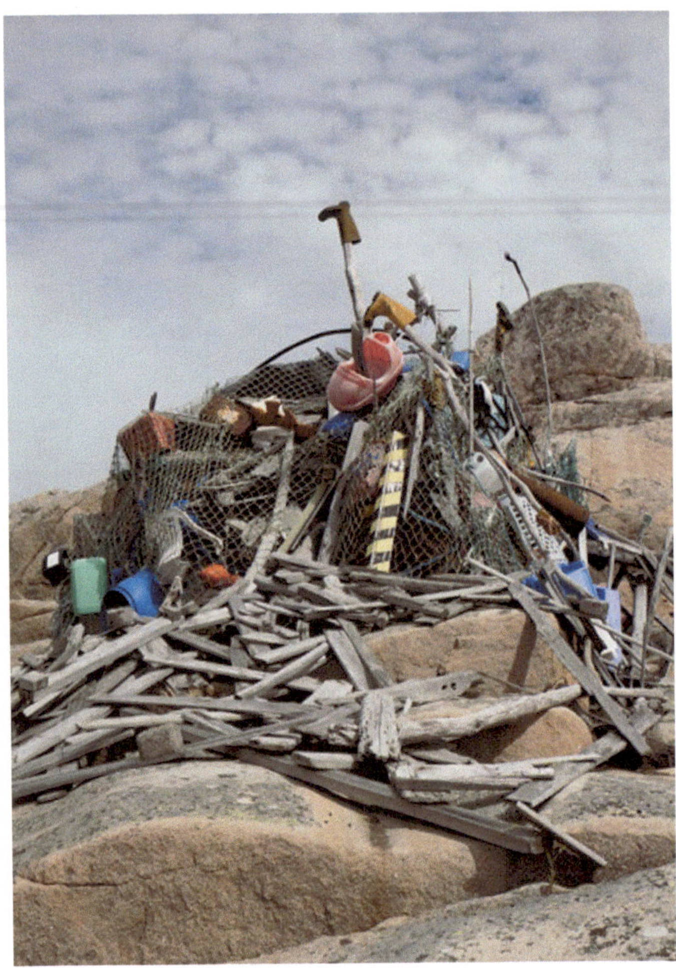

Fig. 7.1 Where does all this come from? *Photo* Hillevi Nagel

References

Svensson U, Omstedt A (1998) Numerical simulation of frazil ice dynamics in the upper layer of the ocean. Cold Regions Sci Technol 28:29–44

World Ocean Review 1 (2010) Living with the oceans: a report on the state of the world's oceans (Hamburg, Tyskland: Maribus gGmbH in cooperation with Future Earth, Kiel Marine Sciences). https://worldoceanreview.com/en/

Chapter 8
To Explore the Unknown

Abstract There are still many gaps in knowledge for every discovery and obstacle. Humans have always sought to expand the learning. We want to know what's behind the horizon and why we live. Exploring the unknown requires curiosity and bravery, feelings within all who wish to learn more. Science and art explore the unknown but use different methods. Open communication and free information flows are fundamental to all attempts to explore the unknown, which applies to science and dreams.

Dreams that come during the night or early in the morning arouse emotions and are often, at first glance, fleeting and difficult to understand. The following chapters present a series of dreams to illustrate how to explore them.

I am out sailing in an old sailing ship.
We are sailing towards a large city, and we are in a hurry—
maybe someone is following us?
We are under full sail, and suddenly, the mainmast is breaking.
On board are the male crew as well as women and children.
The women and children must be rescued first.
Three children are inside the ship, and two are weeping
because flying bats surround them,
some of which are sitting on the children.
One of the children explains that because of the light from above,
the bats like to stay on the boat and are not dangerous.

The dream evokes feelings of freedom, stress, danger, and fear that give way to relief after hearing that the bats are not a threat. The three children inside the old ship are the two life phases: the young child and the aged parent. The bats are dreams, and the light can be seen as the wisdom dreams bring. A new side of me understands that dreams do not mean danger and tries to ease the fear felt by the other two children. The dreamer needs to tell these two about the positive side of dreams. In dream groups, I have often experienced how strongly dreams support us and have

A. Omstedt, *A Philosophical View of the Ocean and Humanity*,
https://doi.org/10.1007/978-3-031-64326-2_8

understood that they are not dangerous, even though they can be violent. By putting words to dreamed feelings and experiences, we can strengthen our mental health and develop more acute perceptions. Inexperienced dreamers often interpret their dreams negatively, while dreams often contain strongly supportive messages that can help dreamers cope with their lives in a new way. A supportive dream group can carefully emphasize the positive messages in dreams. The old ship and the breaking mast symbolize that part of me that needed to find new ways to revitalize. Reconnecting to my younger self supported me in the inner journey.

I am walking towards a market where I see a friend and another man discussing.
I walk towards them and find that my friend has a baby trolley
containing a boy sitting in what looks like a large egg.
I turn towards the child and take him out of the trolley.
We started speaking, and I asked him how he was.
The boy replies that he is happy,
and I feel that it is very important to communicate with him.
He looks ragged and in need of care.

The dream awakens feelings of caring and involvement. The egg-shaped trolley symbolizes the birth of something new, and the boy is a part of myself that I have not yet nurtured. I am now more open to the experience of my own inner experience.

It is early summer, and my family and I are out walking.
A young boy is following us, and I must care for him.
We come to a cafeteria in the park and sit down around a table.
We are speaking together, and I notice that the boy has picked up a flying toy,
which he throws into the air.
It is a beautiful dragonfly that flies around our table, and we are all excited,
although I know that it is just a toy.
The boy then picks up a metal bullet and throws it towards me.
I am afraid but catch it from the air.

The feelings evoked by this dream are joy at being with my family, curiosity when the unknown boy enters, a mixture of concentration and amazement at seeing the airborne dragonfly, withdrawal and perhaps disappointment when realizing that it is just a toy, and fear when the boy throws the bullet. The family symbolizes a well-known and important part of my life, while the boy is a young part of me now calling for my attention. The attack from the boy indicates the importance of attending to my new, growing consciousness.

Two foreign children come to me.
bearing a letter and a cheque for 90,000 pounds.
The sender asks me in the letter.
to take care of these two children.

These dream images awaken feelings of surprise, dedication, and responsibility. The foreign children represent unknown parts of me, and the letter calls for attention. At the same time, the check indicates that doing so will bring great rewards. The number 90,000 is large, and at the same time, it is a phone number formerly used in Sweden to call for emergency services. It was introduced in 1956 and replayed by the number 112 in 1996.

The scientific process is focused on how we can expand the knowledge that is available today. It is thus a risky project with unknown answers and a search for new knowledge. It does not promise simple solutions or products but often leads to further questions at new levels of understanding (Fig. 8.1).

Knowing what controls physical processes such as currents, turbulence, waves, temperature, salinity, and ice formation is fundamental for knowledge of the oceans. You cannot understand what is changing in a marine system without being able to describe these factors, but they do not make up the whole picture. Measurements of oxygen concentrations indicate strong, downward trends in many coastal areas. The reason for the negative trend requires understanding how plankton generates oxygen through photosynthesis, how oxygen from the atmosphere is taken up in surface water, and how dead biological material lowers oxygen levels by breaking down, i.e., being mineralized. Nutrients play an important role in these processes, as they are used during plankton blooms and released during mineralization.

At the same time, the gas composition of the atmosphere changes due to human activity, and the concentrations of carbon dioxide increase. What effects these changes will have on our ecosystems is still mostly unknown and requires new research efforts. The increased eutrophication in coastal seas, such as the Baltic Sea, is partly due to human activities on land—agriculture, land use change, and urbanization—which send increasing nutrient loads and substances to the ocean. But eutrophication also comes from nutrient-rich sediments that release phosphorus when they become anoxic. The causes of eutrophication have been debated by scientists for decades, and research groups have conflicted for years about who is right. At the same time, science has become much more international, for example, in the Baltic Sea region, with the collapse of the Soviet Union and the rise of the European Union. Great progress has been made in promoting a more open scientific attitude. However, scientists are not much different from other people. Like most people, they seek personal and group benefits. The scientific process further includes many self-regulatory mechanisms to correct incorrect information. The most distinguishing characteristic of valid science is reproducibility. If researchers from different research groups cannot reproduce new results, they must conclude that they are invalid. Therefore, open communication and free information flows are fundamental to all attempts to explore the unknown, which also applies to literature, art, and dreams.

Fig. 8.1 What does our relationship with the ocean look like? *Photo* Hillevi Nagel

Chapter 9
Inspiration

Abstract Inspiration can be considered a lighthouse that guides humans and avoids destruction. Today, various environmental threats are reported in mass media without particular focus or context. Instead, lots of competition is taking place in the society. The many challenges society must address are mixed, often without direction. Humanity is lost in uncontrolled development and needs inspiration to generate a common vision, inspiration from our hidden intelligence to find new directions in life.

I am on the Swedish west coast in an older time.
We are staying in a house by the sea,
waiting for some people coming from the sea.
I asked my brother to prepare for the visit.
I see now that the guests are coming—
an old sailing ship with two masts is approaching.

When the ship comes closer, the water level sinks,
and a large stone becomes visible.
The crew sails the ship near the stone.
They throw an electric net over the stone and
blast it into thousands of pieces that,
like a white cloud, disappear in the air.
I tell everyone in the house to stay away.

Many feelings are aroused: anticipation when waiting for the ship, peace, and contentment because of the natural environment, happiness at seeing the vessel, being impressed at the crew's skill, worry when the water level sinks, curiosity when the net is brought out, relief when the stone is blasted, and caution at the end of the dream.

The old house on the Swedish west coast is one of our marine research stations, the Bornö Marine Station (Fig. 9.1). The station was built over a hundred years ago

A. Omstedt, *A Philosophical View of the Ocean and Humanity*,
https://doi.org/10.1007/978-3-031-64326-2_9

and has been an important meeting place for marine scientists, particularly oceanographers. The old ship and historical setting refer to old methods. The stone represents barriers to sound marine management, promoting healthier ocean conditions. The electric net symbolizes modern means of eliminating communication barriers like the Internet. The cloud indicates that I must stay away and wait for clear air.

This dream occurred when I was the scientific coordinator of the Swedish Institute for the Marine Environment (SIME) and was launching a new Internet portal for marine information. Swedish marine management was a mess, with strong competing actors working at various national institutions and universities. SIME leadership was disorganized and harbored different hidden agendas. The stone symbolizes the barriers in the main players' minds. Still, the dream's inspiration suggests that the Internet will blast exclusivity and local competition away, opening up better lines of communication. This dream fostered two attitudes within me: active, challenging barriers by using the Internet, and a more passive one, focusing on avoiding harm. Soon after this dream, I left SIME to concentrate more on science.

I see a woman on a horse.
She bends down and licks away blood from the horse.
I am up in the air looking at the horse
and see three American Indians coming towards the horse.
I understand that this is what is needed to cure the horse.

The dream, which came to me just after I had left SIME, illustrates how my strength and inspiration, represented by the horse, had suffered. The woman symbolizes the need for compassion and healing. The dream also points toward a need for a new perspective and inspiration from a new energy source, represented by the three American Indians. The night after this dream, I dreamt the next dream.

I am up in the Alps and see many horses approaching me.
In the Alps, I am supposed to meet a well-known Swedish climate researcher.
He is not there.
Instead, a well-known Swiss climate scientist meets me
and guides me to a village and restaurant.
Before eating, we climb up on a rock
where I met C.G. Jung.
He takes up a bottle containing crystal-clear water and drinks.
Then he gives me the bottle.

Being high indicates that I am searching for an alternative way, hoping for guidance from my scientific understanding represented by the two climate researchers who have been important persons in my scientific career. Instead, I meet another part of my mind described by C.G. Jung, the Swiss psychotherapist well known for his deep understanding of dreams (Jung 1985). Drinking crystal-clear water with him strongly encouraged me to continue exploring my inner ocean.

Fig. 9.1 Bornö Marine Research Station in Gullmar Fjord has played an important role and inspired many marine researchers. *Photo* Anders Omstedt

I am at the seashore and running along a bridge with my daughter.
I try to jump into the sea but instead stay in the air.
My daughter is sitting on my back,
and now I am flying near the water's surface.
Then we started to fly higher, which gave me a good view of a well-known archipelago along the Baltic Sea coast.
Feeling that I am too high up,
I calm myself, slowly descend towards the water,
and we dive in.
We swim back to the shore,
and I realize that I should not have flown so high with my daughter.

This dream evokes feelings of freedom, danger, and love. Swimming together symbolizes a new relationship between me and my caring feelings.

Today, various environmental threats are reported in mass media without particular focus or context. It is well known that too much nutrients leak into the oceans, leading to eutrophication and oxygen-free bottom waters with dead ecosystems. However, the serious damage that overfishing causes to the marine ecosystem is only known to a small group of scientists. They recommend fishing restrictions to the community, which often ignores them. Climate change and releasing hazardous substances such as plastics and medical residues are ongoing, and the entire marine ecosystem is threatened.

Climate change can cause many problems. A growing concern is that the increased burning of fossil fuels affects the atmosphere's and ocean's temperature and increases the ocean's acidification. Society expects clear and simple communication from researchers. Still, within the research community, there is a growing understanding of the importance of developing tools to manage the combined effects of different effects on the environment. This mix of threats can cause unforeseen and serious damage to ecosystems. Suppose scientists cannot isolate and detect changes because many factors influence them or cannot explain the causes of the changes. How can they then guide society to restore a healthy environment? In climate research, discovering serious changes and what they are due to is called detection and attribution. Similar methods are a prerequisite for improving ocean knowledge and inspire scientists to strengthen their methods today.

Reference

Jung CG (1985) My life. Memories, dreams, thoughts (in Swedish). Nature & culture. Stockholm

Chapter 10
About Our Hidden Intelligence

Abstract Intelligence is the ability to develop thinking, reasoning, planning, solving problems, thinking abstractly, and understanding ideas and language. Intelligence also means access to a large memory capacity, effective searchability, rational understanding, and creative imagination. These characteristics are found in unconscious processes, also called unconscious intelligence. Dreams are part of this intelligence that can enlighten the dreamer about their truth.

When we dream, the dreamer spontaneously creates symbols and metaphors without conscious intention. These form the basis for understanding the dream's message and start from feelings that are generated (triggered) in our daily lives but do not stop there. Certain emotions that are experienced during the day and stay with us and enter the domain of sleep are called day residuals. However, the dream images are not about what triggered the dream image but more about the dreamer's feelings. The symbols and metaphors of the dream are ambiguous and can mainly be understood by the dreamer himself. Montague Ullman spoke of the dream as part of our internal intelligence that can enlighten the dreamer about their truth (Ullman 1996).

How sleep helps consolidate learning and memory is still unknown. Converting the information into images seems to be a good memory strategy. The brain remembers pictures more easily, so it's unsurprising that many claim to remember people's faces better than their names. Dreams are experienced as sequences of mental images, and images convey much more than texts do. Many schools of psychology have developed different thoughts about dreams. Forerunners like Sigmund Freud and Carl G. Jung have inspired many researchers and artists. Ole Vedfelt, an analytical psychologist, has shown that we know more than we think with the help of our subconscious part (Vedfelt 2000). Intelligence is the ability to develop one's thinking, including reasoning, planning, solving problems, thinking abstractly, and understanding ideas and language. Intelligence also means access to a large memory capacity, effective searchability, rational understanding, and creative imagination. According to Vedfelt, these characteristics are found in unconscious processes, which he calls unconscious intelligence. Ove Boudin (1950–) has long been fascinated and thought about his dreams (Boudin 2022). He believes that dreams come from something he

A. Omstedt, *A Philosophical View of the Ocean and Humanity*,
https://doi.org/10.1007/978-3-031-64326-2_10

calls the dream mind, independent of time, is three-dimensional, and stands outside our controlled self. A completely magical and imaginative quality is also the source of our creativity. Others speak of our inner compass or conscience; in the Bible, all dreams are interpreted as divine guidance. An example is the story of Jacob's ladder, which connects the earth and God. Jacob saw this ladder in a dream vision, according to Genesis 28:12. The ladder is a metaphor. It cannot be found in any single place as a physical ladder but is interpreted as a promise of a new opening between humans and God.

The philosopher Karl Popper (1902–1994) stated that we cannot know anything completely; instead, he developed the falsifiability criterion to distinguish science from non-science (Persson 2014). For example, finding a black swan can easily prove that the statement "all swans are white" is false. The falsifiability criterion does not apply to dreams since only the dreamer can interpret the message. Appreciating the importance of understanding dreams and emotions gives an aha experience. It is not a given, true idea but a vision that does not follow logical rules. In our search for ourselves, our inner identity is built up by the aha moment and the memories that appear in our dreams. Ullman (1996) emphasizes that the dream, in its general dynamics, is a natural healing system to support the survival of individuals and species. This healing function is realized through three special abilities of the unconscious, namely the ability to:

- React creatively to something new,
- discern truths that escape us when we are awake,
- and connect with others.

References

Boudin O (2022) Dance under the diamond sky (in Swedish). PianoForte Publishing House, Partille

Persson U (2014) Karl Popper is the prophet of falsification (in Swedish). The Swedish Humanist Association's publication series no. 131. CKM Publishers, Stockholm

Ullman M (1996) Appreciating dreams: A group approach. Sage Publications Inc. International Educational and Professional Publisher. Thousand Oaks

Vedfelt O (2000) Unconscious Intelligence. We know more than we think (in Swedish). Nature & culture. Stockholm

Chapter 11
Summary, Part I

Abstract Here, the first part of the book is summarized. This part introduced the reader to how analytical thinking and intuition can be connected. Here, we use photos, poetry and dreams to examine how fleeting feelings can be translated into stories of great value. Our imagination and our dreams form the basis of all human creation. If society has overconfidence in rationality, we are deprived of the possibility of developing these inner resources. The book's first part aims to increase the understanding of intuition and how we can rationalize and intuitively think.

To free myself from feelings of being frozen and trapped by the expectations of others, I began to explore science, literature, and dreams to find knowledge, meaning, and inspiration. I now sought the source of my feelings—the Nile within me. Seeking knowledge is a journey with no known destination. This is illustrated here by summarizing the chapters above. The dream of my young son climbing the deck and falling into the sea showed me the importance of taking responsibility for my visions and ideas. One must understand that this is the crucial starting point for every person who wants to enter the world of science or learn more about themselves. Steering a boat from the Baltic Sea into the Kattegat, as in the dream in Chap. 4, was an image that visualized my new efforts to change course in life and realize that creativity cannot be controlled by will and demand; instead, it is a gift from our subconscious part.

My scientific work has made it increasingly clear that we live in the Anthropocene, the new epoch when human activity affects the earth's climate and ecosystems. As humans, we still lack the tools to deal with the challenges of this era, and although much scientific knowledge is available, we also need new ways of thinking. Different scientific initiatives seek to engage many different categories of people to tackle the great challenges facing the earth. Hopefully, these initiatives can give us the instruments we lack to create better and sustainable changes. But to develop an improved and sustainable relationship between humans and the ocean, the wealth of intuition, values, and curiosity needs to be considered.

Chapter 8 reproduces a series of dreams that revolve around how to explore the unknown inner ocean with feelings and new insights. In the first dream of an old sailing ship with bats, I realized that dreams are not dangerous but need to be given

A. Omstedt, *A Philosophical View of the Ocean and Humanity*,
https://doi.org/10.1007/978-3-031-64326-2_11

their rightful place. My background and my inner child began to find a voice. The dream of walking with my family evoked a feeling of happiness as I reconnected with my internal ocean. The beauty of a dragonfly and the boy's attack illustrated the need to pay full attention to what is happening. The dream of the foreign children and the 90,000 check reminded me of the need to take care of myself and be aware of how my life had affected my inner life up to that point. In Chap. 9, a series of dreams continues. The inspiring dream of historical barriers and the Internet illustrated a way forward. The injured horse and the need for a new perspective emphasized the need to release frustration and change working conditions. The dream of meeting and drinking water with Carl G. Jung inspired me to continue exploring my internal mind. Finally, the dream of flying with my daughter and reuniting land and sea indicated a new approach without burdens.

The meaning of dreams is to be found in a non-logical understanding. Dreams visualize the communication between outer and inner reality. The inner reality can be seen as the Nile River before the Aswan Dam was built, a rich internal resource that fertilizes our souls. And true freedom is being able to follow your understanding and intuition. In the next part of the book, we will apply what we learned in the first part, mainly through increased knowledge of image interpretation and intuition, when we reflect on how we can change our behavior toward the ocean. Here, a conversation occurs between a worried marine scientist who describes important processes in the sea and possible threats, where the ocean communicates with images. From a philosophical point of view, the dialog is a meeting between science and art or between facts and interpretations (Fig. 11.1).

With her strong longing, Emily Dickinson (1830–1886) inspires us with the following poem:

My River runs to thee—
Blue Sea! Wilt welcome me?
My River wait reply—
Oh Sea—look graciously—
I'll fetch thee Brooks
From spotted nooks—
Say—Sea—Take Me!

(Emily Dickinson, 1976, My River runs to thee).

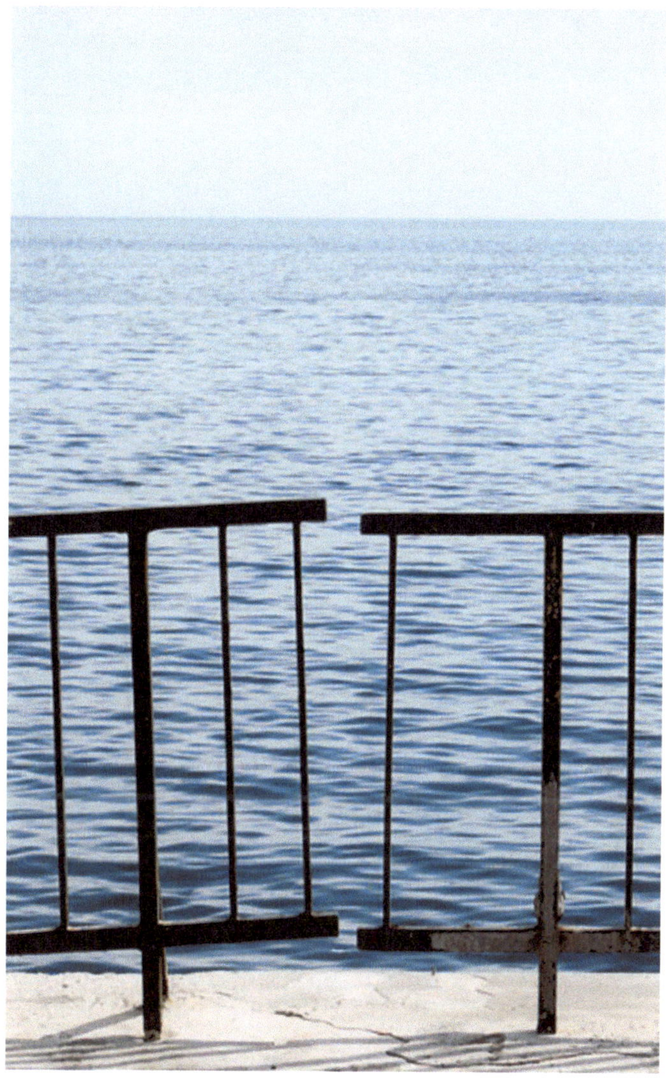

Fig. 11.1 How do we open a new relationship with the ocean? *Photo* Hillevi Nagel

Reference

Dickinson E (1976) My River runs to thee. The complete poems of Emily Dickson. Johnson TH (ed) Poem 162. Back Bay Books, Little, Brown and Company, New York, Boston, London. ISBN-HC-9780316184144. Also available at https://hellopoetry.com/poem/3260/my-river-runs-to-thee/

Part II
In Search of a New Ocean Relationship

Chapter 12
Opening, Part II

Abstract This opening chapter introduces the ocean and its coastal seas' threats from strong anthropogenic pressures. Addressing these threats requires a great change in human behavior, as communicated by scientists and in the United Nations 2030 Agenda for Sustainable Development. To meet this challenge, the science community must work across many academic disciplines using trans-disciplinary approaches and develop new skills for communicating with society. Such communication occurs via drama, scientific and literary writing, and global and regional assessments conveying increasingly urgent information. However, new and more facts alone may not be enough to change behavior. The chapter opens with a discussion of the need to couple science and the arts, with the latter providing knowledge of societal and human values necessary to foster change.

Who asks what the nature of the ocean is?
Who is driven to interpret the law of the ocean?
Those who are driven to ask what their own nature is.
Who am I?
That question can be asked by all.
If you ask the ocean, the ocean will give you answers.
(Sven-Bertil Taube, Who asks the ocean. Taube 2014).

This part concerns the ocean and our common future. It is written from two perspectives: an analytical, scientific perspective and an intuitive, artistic one, where the sea becomes a collaborative partner. The link between human dependence on and feelings toward the sea is examined in a dialogue between these two perspectives. It illustrates how science and art can be connected to increase our awareness of the state of the ocean and support behavioral change. Here, we apply the experience from part one of the book, where analytical thinking is connected to intuition (Omstedt 2020).

Human activities increasingly threaten the ocean and its coastal seas (Reckermann et al. 2022; World Ocean Review 1 2010; World Ocean Review 7 2021). Science struggles to investigate and describe these threats, and to do so requires a broad understanding of various natural science disciplines—including oceanography,

A. Omstedt, *A Philosophical View of the Ocean and Humanity*,
https://doi.org/10.1007/978-3-031-64326-2_12

meteorology, hydrology, geology, geography, chemistry, and biology—but also of human behavior from disciplines such as literature, psychology, history, philosophy, law, sociology, anthropology, political science, and economics. But it is beyond doubt that society must change its behavior and strive to achieve sustainable use and interaction with the ocean. The 2030 Agenda for Sustainable Development, adopted by the UN in 2015, contains 17 Sustainable Development Goals and 169 subgoals related to these challenges. It constitutes an integrated and indivisible vision for the future, intended to balance economic, social, and technological progress in harmony with nature. United Nations has also proclaimed the Decade of Ocean Science for Sustainable Development (2021–2030) as an opportunity for marine scientists and societal actors to show how the human-ocean relationship can be improved (Pendleton et al. 2020).

Ibsen's (1882) play *An Enemy of the People* illustrates how scientific truth comes into conflict with society. The conflict starts with Dr. Stockmann investigating the water in the city spa. After careful investigation, he finds that bacteria seriously pollute the spa water and identifies the source of the contamination—his father-in-law's tannery—and the changes needed to eliminate it. He and his wife are proud and hopeful that he can contribute positively to the community. Soon, various townspeople started to rally around, supporting Dr. Stockmann in criticizing the community leadership and encouraging the local press to write articles on the matter. Dr. Stockmann feels that he has strong support. However, soon, Dr. Stockmann's brother, Peter, who is the mayor, chief of police, and chair of the spa board, visits him, very upset. Why has this investigation been conducted behind his back? Peter argues that water quality is not just a scientific problem but also involves technical and economic considerations. According to the mayor, repairing the spa will be very expensive, requiring money to be raised through increased taxes. Peter asks Dr. Stockmann to announce that his analysis is flawed, but Dr. Stockmann believes he is right and has the people and press behind him. Inspired by the truth, he writes an expose for the local newspaper and gets support to publish it. Peter contacts the newspaper and argues that the claims about the spa will ruin the community, and support for Dr. Stockmann starts to ebb. Instead of Stockmann's article, the local newspaper published a short announcement from the mayor that there was no problem with the spa. Dr. Stockmann's next approach is to give a public meeting. After some delay, Dr. Stockmann started his presentation, but, having lost focus, he presented his insights into how society works instead. Inspired, he states that the community's spiritual life is completely poisoned. The audience turns against him; Stockmann is called an enemy of the people, and the meeting breaks up. By the next day, the windows of Stockmann's home have been broken by stones thrown by upset townspeople. The mayor, his brother, gives him a letter stating that he has been fired from his job at the spa. Stockmann's daughter has lost her job as a teacher, and the family has been evicted from their house. As an enemy of the people, all of society has turned against him. Still, he decides to stay and fight, even though the mayor and others recommend leaving, and he later admits that he has made a mistake

Could Dr. Stockmann have taken a different route to improve the water conditions in the spa? The play indicates he made a mistake by going behind the mayor's back. Perhaps another alternative would have been to wait for people to become sick from the spa water. We don't know what happened to Dr. Stockmann and his family. However, Ibsen's drama about a whistleblower is still relevant today when scientists or other experts try to communicate new results that conflict with society's beliefs or economic and social interests

In 1962, marine biologist Rachel Carson published her famous book *Silent Spring* about the increasing danger of using chemicals like DDT to control insects (Carson 1962). After publishing the book, Carson endured attacks from the agricultural industry and its government allies, who tried to deny her message. Despite well-funded and bitter personal attacks, she spread her warning about chemical pollution. Few books have had such an impact on public awareness of environmental degradation. In an afterword to the 1999 edition of *Silent Spring*, Linda Lear noted that despite Carson's struggle, the world now finds itself awash with thousands of new and dangerous chemicals at the beginning of the new millennium. The problems with plastic in the ocean and pharmaceutical residues in the wastewater have not yet been clarified.

In biologist Alexandre Antonelli's (1978–) book *The Hidden Universe. Adventures in Biodiversity,* he estimates that humans produce more than 350,000 different kinds of chemicals and chemical mixtures (Antonelli 2022). Sooner or later, many of these end up in nature and the ocean, often with unknown effects. Antonelli describes how the diversity of biological life on earth is a gigantic universe that is threatened in many ways.

In his book *The Deep: The Hidden Wonders of Our Oceans and how we can protect them*, marine biologist Alex Rogers makes a strong plea for protecting the deep seabeds (Rogers 2019). Rogers notes that we are now at a critical historical juncture when we have realized that much of the ocean is dying. We face a choice between two very different oceans: a healthy and productive ocean managed sustainably or an overexploited ocean that is increasingly destroyed and where biodiversity is at risk. The former requires, according to Rogers, that all nations immediately introduce an effective agreement to:

- Reduce carbon dioxide emissions,
- eliminate overfishing and destructive fishing methods,
- establish a global network of marine protected areas,
- effectively reduce marine pollution,
- improve marine management,
- map the sea better,
- better map the marine ecosystems and how the ocean works,

- and intensify education about the ocean and its importance.

The researchers' presentations of their results have been improved through international evaluations. The Intergovernmental Panel on Climate Change (IPCC) conducts regular assessments of humans' impact on the climate, thereby significantly enhancing communication between scientists and the rest of society. Similarly, assessments of climate change in the Arctic (ACIA, Arctic Climate Impact Assessment) and the Baltic Sea (BACC, Assessment of Climate Change for the Baltic Basin) have created platforms where researchers and stakeholders can meet and improve communication at a regional level.

Understanding humans' influence on the climate is now well known in society. However, there is strong media competition in today's society, and many compete for attention in the triangle between politics, media, and science. Societal problems related to climate change, eutrophication, alien species, acidification, emissions, plastic waste, and various toxins (for example, arsenic released by industry and pharmaceutical residues from sewage treatment plants) have taught scientists to also compete with each other for the spotlight.

There exists also a clear communication barrier between what researchers say and what the public understands, and terms can have different meanings. For example, "uncertainty" and "positive trend" are interpreted by the public as "ignorance" and "good development" respectively. Another example is the often vaguely defined word "sustainability." The sustainability concept comes from forestry, which means that resources should be used deliberately so that the natural resource never runs out. Today, however, the word is used as an ill-defined slogan in many contexts, and researchers are trying to formulate concrete guidelines for a sustainable future.

Emotions are considered by many to be irrational, and researchers try to present their results objectively, following different strategies without personal values. But communication that only conveys more and more facts is not enough to achieve emission reductions, for example. Something else is needed to add emotional resonance, connect the messages with specific locations, and promote action. Humanities and art forms such as visual arts, dance, theater, literature, and film are forms of expression that can revitalize science with emotion, promote changes in attitudes, and lead to changed lifestyles. Presenting marine science without connecting it to human or societal values can weaken our emotional connection to the ocean.

References

Antonelli A (2022) A hidden universe. Earth's unknown biological diversity (in Swedish). Nature & Culture. Stockholm.

Carson R. (1962/2000) Silent spring. Afterword by Linda Lear. Pe, London, UK

Ibsen H (1882) An enemy of the people. Available at libraries

Omstedt A (2020) A philosophical view of the ocean and humanity. Springer Nature, Cham, Switzerland

Pendleton L, Evanse K, Visbeck M (2020) We need a global movement to transform ocean science for a better world. PNAS 117(18):9652–9655

Reckermann M, Omstedt A, Soomere T, Aigars J, Akhtar N et al (2022) Human impacts and their interactions in the Baltic Sea region. Earth Syst Dyn 13:1–80

Rogers A (2019) The deep: the hidden wonders of our oceans and how we can protect them. Wildfire, Headline Publishing Group, London, UK

Taube SB (2014) Song: who asks the sea (in Swedish), with Olle Adolphson and Peter Nordahl from their album "Hommage"

World Ocean Review (2010) World ocean review 1. Living with the oceans: a report on the state of the world's oceans. Maribus gGmbH in cooperation with Future Earth, Kiel Marine Sciences, Hamburg. Available at https://worldoceanreview.com/en/

World Ocean Review (2021) World ocean review 7. The ocean guarantor of life—–sustainable use, effective protection. Maribus gGmbH in cooperation with future Earth, Kiel Marine Sciences, Hamburg. Available at https://worldoceanreview.com/en/

Chapter 13
An Attempt to Connect to the Ocean

Abstract In this chapter, the connection with the ocean is made through a dream image of a dead seabird that had ingested plastics. The senseless killing of a seabird is the theme of *The Rime of the Ancient Mariner* by Coleridge, illustrating how killing an albatross brought isolation and death in life. The image of the dead seabird can be interpreted as a metaphor for our dysfunctional state of mind, with environmental problems being seen as mental problems. The chapter concludes that humans need help to think more broadly about the ocean.

Voice of the ocean: *Do you think I can harbor all the rubbish you put into me without destroying our relationship? You need to broaden your thinking.*

The ocean, I feel so close to you and am sorry for how we treat you (Eriksen et al. 2023; Reed 2015). People stress you in many ways, and new threats constantly join the ones we already know. Unchecked, humans have developed effective fishing methods that destroy large parts of marine ecosystems. Large areas of seagrass, which play such an important role in marine ecosystems, are severely degraded. People build along the beaches and change your coastlines, spreading nutrients and toxic substances into you, believing you can absorb it all. Today, humans are unpredictably changing the climate through massive carbon dioxide emissions and other greenhouse gases.

You say I need to broaden my thinking, but how? Sorry, I don't know what you mean, as I only saw the messages you sent. People are increasingly disconnected from you, and urbanization exacerbates this, filling our minds with the notion that we are independent of you. Several initiatives are underway to save you, but not all of your coasts. Scientists are trying to understand how the environment and people affect you. But even if people improve their understanding, you always seem to be the loser. Can you give us a hint as to the cause of our extreme environmental abuse?

Thank you for the dream image you, the ocean, sent me of a seabird dying from plastic ingestion (Fig. 13.1). The image is incredibly sad, and I immediately felt sad and angry when I woke up. It reminded me of Coleridge's (1798) poem, *The*

A. Omstedt, *A Philosophical View of the Ocean and Humanity*,
https://doi.org/10.1007/978-3-031-64326-2_13

Rime of the Ancient Mariner, about how the mariner futilely shoots an albatross, how their ship suffers hardship and death, and how, as the sole survivor, he must recount his crime over and over again. I realize that marine ecosystems are not managed as they deserve. Ecosystem services are now spoken of as something that can be calculated and whose value can be estimated. I wonder what services people provide for you? But this may not be what you, the ocean, are trying to say. Perhaps the dream image of the dead bird should instead be understood as a metaphor for our dysfunctional behavior, where we lack the inspiration or value needed to serve the environment. Maybe marine problems are mental problems, and humans need to understand themselves better before they can restore the environment. Please let me know what is missing from our feelings and values so I can connect better with you and promote changes!

....

The ice was here, the ice was there,
The ice was all around:
It cracked and growled, and roared and howled,
Like noises in a swound!

At length did cross an Albatross,
Thorough the fog it came;
As if it had been a Christian soul,
We hailed it in God's name.

....

God save thee, ancient Mariner!
From the fiends, that plague thee thus!—
Why look'st thou so?'—With my cross-bow
I shot the Albatross.

....

And I had done a hellish thing,
And it would work 'em woe:
For all averred, I had killed the bird
That made the breeze to blow.
Ah wretch! said they, the bird to slay,
That made the breeze to blow!

(Samuel Taylor Coleridge, The Rime of the Ancient Mariner, Coleridge 1798)

Fig. 13.1 Ocean sends an image of a dead seabird stuffed with plastic. *Illustration* Jan Heuschele

References

Coleridge ST (1798) The rime of the Ancient Mariner was written by the English poet Samuel Taylor Coleridge in 1797–1798 and published in 1798 in the first edition of Lyrical Ballads, which also contained poems by William Wordsworth. It is found in several translations and available on the link: www.poetryfoundation.org/poems/43997/the-rime-of-the-ancient-mariner-text-of-1834

Eriksen M, Cowger W, Erdle LM, Coffin S, Villarrubia-Gómez P et al (2023) A growing plastic smog, now estimated to be over 170 trillion plastic particles afloat in the world's oceans—urgent solutions required. PLoS ONE 18(3):e0281596. https://doi.org/10.1371/journal.pone.0281596

Reed C (2015) Dawn of the plasticene. New Sci 28–32

Chapter 14
Sea Ice and Curiosity

Abstract Part II aims to establish communication between marine science and the arts by giving the ocean a voice. The first topic is the dynamics and thermodynamics of sea ice and related threats due to global warming. When seen from the geophysical scale, sea ice is a beautiful and fragile plastic–viscous material sensitive to changes in heat fluxes, snow, and wind. It protects the underlying water and serves as a medium for life for many marine species, from ice algae to polar bears. Should humans be fearful, their souls frozen at the ultimately violent use of fossil fuels that has caused the melting of sea ice? The ocean responds by challenging us to cultivate our curiosity. The conclusion is that curiosity is a much better attitude toward the ocean than either fear or guilt.

Voice of the ocean: *Look into me in new, imaginative ways and see the many beautiful species that not only survive but also play beneath the ice.*

The ocean and marine ecosystems are protected by ice formation. Ice changes its relationship with the atmosphere and forms a new interface that affects the exchange of motion, heat, and gases. When the surface water freezes, the water molecules change dramatically, becoming lighter than seawater and preventing deep water from freezing (Wadhams 2000). The salt of the seawater leaves the ice, making the ice crystals consist of fresh water. On a calm water surface, star-like ice crystals quickly grow and overlap, freeze together, and form a thin layer of ice. This type of ice can be easily bent by waves and movement, creating a pattern known as Finger Rafting. During further ice formation, the ice crystals grow to form Columnar Ice.

Ice can form in different ways when winds are blowing over the water. Due to turbulence, the small, initially ice crystals are mixed down in the super-cooled surface layer, forming frazil ice with a typical morphology of plates or disks with a millimeter diameter (Martin 1981). As ice production continues, Frazil Ice forms a layer on the surface called Grease Ice (Omstedt 1985), which is the raw material of other ice types, such as Pancake Ice and Sea Ice.

Many different forms of ice can be found in the freezing ocean, ranging from Columnar Ice, Snow Ice, and Frazil Ice to Ridged Ice, forming a complex environment

A. Omstedt, *A Philosophical View of the Ocean and Humanity*,
https://doi.org/10.1007/978-3-031-64326-2_14

that can interact with marine life in many ways. Sea ice can hinder and threaten shipping and people living in freezing water areas; for others, ice offers a way to travel rapidly and is a good platform for fishing and hunting (Leppäranta 2011).

People who live along ice-covered waters have long known that the ice moves. At the end of the nineteenth century, the zoologist and oceanographer Fridtjof Nansen (1861–1930) explored ice drift in the Arctic Ocean by letting his ship Fram freeze into the ice. The Fram Expedition (1893–1896) was an attempt to use the movement of the ice for travel to the North Pole. The expedition did not succeed in reaching the pole but made many important observations. One observation was that the ice does not drift along the wind but deviates 20–40 degrees to the right of the direction of the wind. Nansen explained this deviation as a result of the earth's rotation. At Nansen's suggestion, the physicist and oceanographer Walfrid Ekman (1874–1954) investigated these observations mathematically. Ekman(1905) presented a mathematical formula explaining how surface currents in the Northern Hemisphere deviate to the wind's right and rotate clockwise down deeper layers. This work formed the basis of modern theoretical oceanography. Later, it was possible to show through current measurements anchored on drifting sea ice that this theoretical prediction could be observed.

Growth and melting of sea ice are closely related to the exchange of heat between the atmosphere, the ice, and the water. The temperatures above the ice surface and solar radiation are the main driving forces from the atmosphere. Under the ice, water warmer than the freezing point is the main cause of melting. Since the melting temperature of the ice depends on the salinity, the temperature at the interface between the ice and the water depends on processes very close to the ice surface (Svensson and Omstedt 1990). Mathematical models of sea ice thermodynamics have a long history, beginning with investigations of heat conduction through ice and snow. The first numerical model of sea ice thermodynamics was developed in the early 1970s. It shows that the thickness of the ice is sensitive to temperature changes (Maykut and Untersteiner 1971).

Significant military tensions in the twentieth century increased the need for national control and surveillance of the cold seas. Through major international programs, drifting buoys were placed on the ice to understand the movement of the ice better, and a long discussion started about how to classify sea ice as a material. This discussion created a foundation for new mathematical modeling and tests of whether sea ice should be modeled as a plastic, elastic, or viscous material. We must now put on our "geophysical glasses" and view the ocean from a helicopter perspective. An examination of satellite images shows that sea ice is a fragile material that easily breaks into ice ridges and leads. Disintegrated sea ice can never return to smooth ice, and it is clear that the sea ice behaves as a plastic material on a geophysical scale. We can also imagine sea ice acting like a viscous material due to internal friction. Nansen discovered internal friction in sea ice fields as an important aspect of ice drift. The ideas about sea ice's internal friction and plastic behavior form the basis of the geophysical sea ice models used in most climate studies today.

Advances in the knowledge of sea ice have required a new geophysical perspective because the laboratory perspective can only examine properties on a smaller scale.

These two perspectives, the global and the local, respectively, are, of course, both valid, but different problems require different perspectives. However, applying only one approach can cause problems when other aspects are ignored or simplified. Over the past decade, the Arctic ice sheet has shrunk significantly during the summer. The reduction in summer ice has also occurred faster than climate model projections have predicted, indicating that something is missing in our understanding. How the melting occurs under the Antarctic ice sheets is still unclear, but the fact remains: glaciers are melting, and rising sea levels threaten coastal areas and islands.

The ocean, I converse with you on a night with a full moon, hoping for intuitive guidance through my inner imager: Should we blame ourselves for using fossil fuels? You know that these fuels have been the basis for strong industrial development. Most of the energy consumed today still comes from oil and coal, which has enabled population growth and improved transport across the ocean.

Thank you, ocean, for your quick response and sending such a nice dream image of snow-covered sea ice (Fig. 14.1). The image is complex, with many beautiful details, and I ask myself what emotions it evokes. Sea ice piques my curiosity and makes me feel alive. Algae can find safe living conditions within pockets in the ice and wait for the sun to melt the ice. It is as if the ice has created a nursery for life. Sea ice is also a good place for seal pups to grow up and be fed. The most beautiful frost flowers form on the ice surface, and entire ecosystems develop in and under the ice. Sea ice ensures the survival of many marine species.

Are you implying, ocean, that the melting is due to warmer atmospheric temperatures and that you are trying to save the ice through various stabilizing effects? I understand you are trying to calm me down and strengthen my curiosity. Curiosity creates a beautiful feeling because it inspires us to explore beyond our limitations, like scientists grappling with unsolved questions, children exploring and playing, or whatever this mental flow of concentration and joy should be called. The alternative is fear and shame, where we ignore pressing challenges and remain paralyzed. I see your point—that curiosity is a much better approach to climate change and that you expect people to change their attitudes and start working with you.

Fig. 14.1 Ocean sends an image that illustrates the complex structure of sea ice. *Illustration* Jan Heuschele

References

Ekman VW (1905) On the influence of Earth's rotation on ocean-current. Arkiv För Matematik, Astronomi och Fysik 2(11)

Leppäranta M (2011) The drift of sea ice. Springer-Praxis Books in Geophysical Sciences, Springer Heidelberg Dordrecht London New York. https://doi.org/10.1007/978-3-642-04683-4

Martin S (1981) Frazil ice in rivers and oceans. Annu Rev Fluid Mech 13(1):379–397

Maykut GA, Untersteiner N (1971) Some results from a time-dependent, thermodynamic model of sea ice. J Geophys Res 76(6):1550–1575

Omstedt A (1985) Modelling frazil ice and grease ice formation in the upper layers of the ocean. Cold Reg Sci Technol 11:87–98

Svensson U, Omstedt A (1990) A mathematical model of the ocean boundary layer under drifting melting ice. J Phys Oceanogr 20(2):161–171

Wadhams P (2000) Ice in the ocean. Gordon and Breach Science Publishers. Amsterdam, the Netherlands. ISBN 90-5699-296-1

References

Chapter 15
Stratification, Turbulence, and Service

Abstract This chapter continues the dialogue between science and the ocean, starting with a concerned science view. A great fear is that global warming may reduce areas of sinking water, reducing ocean ventilation. The ocean responds with an image of a kelp forest—among the most beautiful and biologically productive marine ecosystems and one that has served the ocean for many millions of years. Interpreting the kelp forest as a metaphor suggests a need to listen better and realize that humans can similarly develop society as a rich habitat for human growth with diverse groups of people in a healthy environment. For this to happen, human attitudes must change to create a society that serves the ocean like kelp forests do.

Voice of the ocean: *It is such a joy to serve life. Some of you speak of "green thumbs," but I have "brown fingers" that have protected and served life for millions of years. Listen carefully.*

So much in the ocean is about gravity and density. Gravity is an invisible force that pulls bodies together. It is easy to comprehend the great importance of the sun for the ocean. Without the sun, the first thing that would happen is darkness. The gravitational force from the sun would be gone, and the earth would start to move out into space. The temperature would drop dramatically, plant photosynthesis would stop, and the earth would become completely covered with very thick ice. Now, what about the moon? Without the moon, the nights would be much darker. The days and nights would be much shorter as the moon slows down the earth's rotation.

Gravity also plays an important role in fluid motion (Cushman-Roisin and Beckers 2011). In the ocean, it stabilizes or destabilizes water of different densities, with density being a measure of mass that can be calculated in this context from observations of salinity and temperature. The water becomes stable if light water is above dense water and unstable if dense water is above light water, forming stratified and mixed layers within the ocean. In some important surface areas, the surface water becomes denser than the underlying layers due to water cooling and, during ice formation, through salt rejection. These areas in the polar oceans ventilate the deeper ocean by introducing oxygen-rich surface water to deeper layers. Sinking of surface water

A. Omstedt, *A Philosophical View of the Ocean and Humanity*,
https://doi.org/10.1007/978-3-031-64326-2_15

also occurs in coastal seas when, for example, cold winter winds blow over the Bay of Lyon in the Mediterranean.

The ocean is stratified with layers of various densities that, through gravity, arrange themselves so that lower densities are found above higher densities. Motion in the water continually disturbs this stratification and, if strong enough, diminishes it. Our early understanding of circulation between surface and deeper water layers was based on temperature and salinity observations. Looking into the ocean, one can find different water types coming from different areas, such as Arctic Ocean surface water in the deep layers of the Atlantic Ocean, indirectly illustrating the ocean circulation. Different water masses can be transported long distances, virtually throughout the world's oceans. Surface water from the polar oceans flows into the deeper ocean layers, just as, for example, surface water from the Kattegat enters into the deeper layers of the Baltic Sea.

Similarly, dense water from the Mediterranean Sea spills over the Gibraltar Strait into the Atlantic Ocean and sinks into the deeper layers there. The density differences on the vertical and horizontal axes make environmental and biological conditions quite stable in the oceans. Many things will change if the ocean becomes warmer due to climate change. The sea level will rise, sea ice and glacial ice may melt, and warmer water will profoundly influence the marine ecosystem. A major threat to the ocean will be if deep water ventilation is reduced, as the dense surface water injects oxygen into the deeper layers.

Simultaneously, the water is mixed by turbulent eddies. Many research efforts have been dedicated to understanding turbulence, and much has been learned from laboratory studies. The energy to feed turbulence comes from shear currents, convection, and breaking waves, which make the flow unstable, chaotic, and random, including a whole range of eddies of various types. At the surface, turbulence is driven by winds and by changes in salt and heat content. In deeper waters, turbulence is caused by astronomical factors, such as the moon and sun, generating internal waves that break against bottom topography (Sjöberg and Stigebrandt 1992; Vic et al. 2019). This means deep water mixing depends not on meteorological forcing but on bottom structure and astronomical forcing. Deep water mixing reduces vertical stratification and facilitates ventilation by overturning the surface water. This may mean that the dense surface water, even if it becomes warmer, may still be able to ventilate the deeper layers as long as certain factors remain unchanged, factors such as the bottom topography, with its ridges and canyons, and astronomical forces such as the moon's orbit. Could you, the ocean, add anything about this? Will climate warming make your deep water anoxic, or will tides generated by the moon help?

I had been eagerly waiting for almost a week for your reply, and then tonight, I got a message (Fig. 15.1). It isn't easy to interpret, as you only sent me a dream image of large brown seaweeds—kelp. The image is beautiful, and I am gaining respect for this marine species growing in the ocean for millions of years. Interpreting the image as a metaphor reminds me of the freedom of swaying in the water currents while my soul is connected to the foundation from which everything grows. Perhaps you are trying to tell me to be more patient and learn to listen. Yes, there is a lot of noise in modern society, and many voices compete for attention. Listening is not the same as hearing and is not easy, but calling for attention to what is communicated and keeping an open mind. This internal image, apparently of kelp, conveys something quite different from other messages I hear, many of which are about environmental problems that seem almost beyond human remedy, as if civilization were about to collapse. Kelp can develop dense forests with high biodiversity and many ecological functions. They may be among the most beautiful and biologically productive habitats in the marine environment. Aha! You are signaling that people can develop society similarly, as a rich habitat for human growth with diverse groups of people in a healthy environment. I will try to listen more carefully and think more about how to serve the ocean sustainably, as the kelp forest does. I know that what one pays attention to tends to grow in importance in one's mind, so I need to be more patient.

Fig. 15.1 Ocean sends an image of a marine forest. *Illustration* Jan Heuschele

References

Cushman-Roisin B, Beckers J-M (2011) Introduction to geophysical fluid dynamics: physical and numerical aspects. In: International geophysics series, vol 101. Academic Press–Elsevier, Waltham, MA. ISBN 987-0-12-088759-0

Sjöberg B, Stigebrandt A (1992) Computations of the geographical distribution of the energy flux to mixing processes via internal tides and the associated vertical circulation in the ocean. Deep Sea Res Part A. Oceanogr Res Pap 39(2):269–291

Vic C, Naveira Garabato AC, Green JAM, Waterhouse AF, Zhao Z et al. (2019) Deep-ocean mixing driven by small-scale internal tides. Nat Commun 10:2099

Chapter 16
Currents and Vulnerability

Abstract The ocean is never at rest. Currents, eddies, and turbulence change at different time scales in a beautiful and complex manner. Human misbehavior is clearly seen in the plastic waste from various sources transported all over the ocean by these currents. Some believe that if we do not change our behavior, the ocean will eventually contain more plastic than fish. Instead of pristine ocean eddies, plastic will be accumulated by the currents into large surface "garbage patches" that are very dangerous for sea birds and marine ecosystems. Without a better human connection, the ocean will become a biological desert mirroring our alienation and frozen emotions. The ocean's response to this misbehavior reminds us of our wasteful lifestyles and the need to improve our mental health. Both the ocean and humans are vulnerable, and we humans need inspiration to change our behavior.

Voice of the ocean: *I sent you an image of a dying seabird with plastic in its stomach as a reminder of your mental health. You need to feed your internal ocean in a better way. No one can survive without inspiration; I can be your strong inspiration.*

Ocean water always moves through waves, currents, eddies, and turbulence. Sea levels rise and fall through tides while currents form along the coasts. On a larger scale, the surface currents are driven by winds, which give rise to ocean circulation. The strong westerly winds over the Atlantic generate surface water transport to the south—called Ekman transport after the physicist and oceanographer Ekman (1905). At the same time, the trade winds generate an Ekman transport to the north. These two surface transports pressure the underlying currents, forcing large water masses to move slowly toward the equator. In the Southern Hemisphere, the winds, again through Ekman transport, drive a slow current toward the equator. This transport is called the Sverdrup transport after Harald Sverdrup (1923–1992), who published his discovery in 1947 (Sverdrup 1947). The water transported to the equator must somehow return to the polar regions, and the only possibility for this is via fast western boundary currents, such as the Gulf Stream in the North Atlantic, the Brazil Current in the South Atlantic, and the Kuroshio Current in the North Pacific. These currents

A. Omstedt, *A Philosophical View of the Ocean and Humanity*,
https://doi.org/10.1007/978-3-031-64326-2_16

transport large amounts of water, forming eddies and home to marine ecosystems (Munk 1950; Stommel 1948).

The largest current system is around Antarctica, which runs clockwise around the continent and is forced by strong westerly winds. This current system protects the Antarctic ice sheet from warm water, making westward sailing difficult. In the Arctic, the situation is different. Water enters from the Pacific through the Bering Strait and flows through the Arctic out into the western Atlantic. The flow is forced by the difference in sea level between the Pacific and the Atlantic, as the Atlantic has higher salinity and lower sea level than the Pacific. The National Aeronautics and Space Administration (NASA) has made beautiful films illustrating the large-scale multi-eddy currents that form in the ocean.

But it is not only tides and winds that drive currents but also changes in the density of seawater. On a global scale, we talk about thermohaline circulation also called the Great Ocean Conveyer Belt, where heavy surface water sinks to deeper layers and eventually rises to the surface, but now in a completely different area (Broecker 1991). Sinking heavy bottom water from Antarctica follows the continental shelf. Detailed depth measurements show many channels, valleys, and other small-scale local topographic features influencing and channeling the water flow (Muench et al. 2009). Changes in temperature and salinity generate currents in most coastal oceans, where even rivers and precipitation can change the circulation, with the Baltic Sea and the Mediterranean being good examples.

Wind, temperature, and salinity generate areas of vertical circulation. Rising water (upwelling) near the coast can bring nutrients from deeper water layers to the surface, supplying plankton with nutrition. The opposite can also occur when the surface water sinks (downwelling) and reduces the supply of nutrients to the surface, inhibiting the plankton bloom. Ocean currents and circulation serve humanity by facilitating sea transport and regulating the earth's climate. For example, the western boundary currents transport warm water toward the pole, making the northern European climate much warmer than at corresponding latitudes along the western coasts of the Atlantic.

The modern era of oceanography began in the late nineteenth century with expeditions by research ships making various marine observations. The Challenger expedition led by Charles Wyville Thomson (1830–1882), which circumnavigated the globe between 1872 and 1876, was particularly important. The ship was equipped with a laboratory and collected many instruments for ocean characteristics, including ocean temperatures, seawater chemistry, currents, marine life, and seabed geology.

Recently, there has been a rapid development of new measurement methods (Wuench 2015), for example, using:

- Satellites,
- free-floating floats,
- bottom pressure gauge,
- and sensors mounted on diving animals, for example, elephant seals.

Marine measurements are still expensive and difficult to make, and the ocean is insufficiently measured. Plastic in the sea is a huge problem for the ecosystems (Barnes et al. 2009), in addition to ice polluted by sediment and land-based materials, oil spills, toxic substances, and garbage. Scientists fear that by 2050, there will be more plastic than fish in the ocean. First, we humans destroyed large fish stocks through extensive overfishing, and now we are filling the ocean with plastic and trash made from oil, natural gas, and coal—pure criminal behavior! Ocean currents transport this plastic worldwide and collect it in large artificial islands, threatening seabirds and other marine life. The mass production of plastic began in the 1950s, and many now realize plastic is destructive to both land and sea.

Is there no end to human stupidity? Why can't we behave more cautiously? In the past, glass bottles, metal cans, and paper bags were used, but then suddenly, we started using huge amounts of plastic without thinking about how to deal with the resulting waste. Why can't people be more inspired by the ocean and learn to manage it better? Otherwise, the ocean risks becoming a biological desert that only serves to illustrate our alienation and our frozen emotions. You, the ocean, seem very quiet, and I wonder if you can say anything more about our destructive behavior.

I feel your presence, but can you come a little closer? Now we are in contact, and I can almost guess that the image you sent to my inner sea can be compared to a lobster (Fig. 16.1). Lobsters are found in all oceans and live on rocky, sandy, or muddy bottoms. They have had several hundred million years to develop. The dream picture evokes feelings of surprise and joy in me. The lobster with its strong claws reminds me of a dream where I was stung by a scorpion on the heel tendon. After that dream, I realized that my vulnerability was the key to my inner life. So maybe you, the ocean, are saying that awareness of plastic can enable us to learn more about the vulnerability of the ocean and humans.

Fig. 16.1 Ocean sends an image of a lobster hidden under a rock. *Illustration* Jan Heuschele

References

Barnes DKA, Galgani F, Thompson RC, Barlaz M (2009) Accumulation and fragmentation of plastic debris in global environments. Philosoph Trans Royal Soc B—Bio Sci 364(1526):1985–1998

Broecker WS (1991) The great ocean conveyor. Oceanography 4(2):79–89

Ekman VW (1905) On the influence of Earth's rotation on ocean-current. Arkiv För Matematik, Astronomi och Fysik. Band 2 11

Muench RD, Wåhlin AK, Özgökmen TM, Hallberg R, Padman L (2009) Impacts of bottom corrugations on a dense Antarctic outflow: NW Ross Sea. Geophys Res Lett 36(23):L23607

Munk WH (1950) On the wind-driven ocean circulation. J Meteorol 7(2):79–93

Stommel H (1948) The westward intensification of wind-driven ocean currents. Trans Am Geophys Union 29(2):202–206

Sverdrup HU (1947) Wind-driven currents in a baroclinic ocean, with application to the equatorial currents on eastern Pacific. Proc Nat Acad Sci USA 33:318–326

Wuench C (2015) Modern observational physical oceanography: understanding the global ocean. Princeton University Press, Princeton, NJ

References

Chapter 17
Heat Balance, Temperature, and Interpretations

Abstract The heat balance that determines the earth's temperature represents an interesting and complex interplay between the sun, the earth, and our behavior. The ocean plays many important roles in this, such as being the earth's most important heat storage sink. Evaporation from the ocean surface due to latent heat flux works like a steam engine, driving large-scale atmospheric circulation. Long-wave radiation emitted from the earth's surface is partly reflected back from the atmosphere by the GHGs that blanket the earth, protecting it from cooling. The human impact comes from anthropogenic landscape change and increases in atmospheric GHG levels, for example, from fossil fuel burning, which influences the long-wave radiation reflected back to the surface. Today, it is frighteningly clear that humans are influencing the ocean through global warming and ocean acidification. Will humans be able to reduce global warming or not? The ocean sends an image of a red jellyfish swimming slowly to the surface, illustrating how humans' internal resources, such as intuition, dreams, and emotions, can foster a more accurate perception of our relationship with the ocean, in all its beauty.

Voice of the ocean: *Humankind thinks in dysfunctional, destructive ways. All of you often think in the wrong direction. I promise I will not seek revenge for your misbehavior. You are your worst enemies, but I will strongly support you however I can.*

The earth's heat balance illustrates the strong connection between the ocean and the sun, even though they are almost 150 million kilometers apart. The distance between the earth and the sun constantly changes in many ways. The earth rotates one revolution in 24 h, with nights followed by bright days. Earth takes one year to orbit the sun, with seasons differing markedly in solar radiation. The earth's daily rotation and annual orbit are counterclockwise, and the earth moves around the sun in an ellipse. As the earth moves around the sun, its tilted axis means that maximum solar radiation hits 23° north of the equator in June and 23° south of the equator in December, at what is known as the Tropics of Cancer and Capricorn, respectively.

A. Omstedt, *A Philosophical View of the Ocean and Humanity*,
https://doi.org/10.1007/978-3-031-64326-2_17

The sun's influence on the ocean is not simple because surface water in different regions comes into contact with the sun's radiation at other times of the year. In addition, the shape of the earth's orbit around the sun varies from nearly circular to slightly elliptical, while the tilt of the earth's axis changes. Some of these orbital changes cause major climate variations. Observations from glacial ice indicate a one-hundred-thousand-year cycle of oscillations between glacial and interglacial periods. These climate variations, estimated to have varied between average temperatures of −8–+ 2 °C in Antarctica over the past eight hundred thousand years, are also documented in other measurements, such as atmospheric carbon dioxide and sea level (IPCC 2013, p. 400). Nevertheless, the sun alone cannot explain this strong centennial temperature cycle, suggesting that internal climate feedback mechanisms are also significant (Abe-Ouchi et al. 2013).

Besides solar radiation, the most important aspect of the heat balance is the latent heat flux associated with evaporation from the sea surface. Through evaporation, the ocean acts as a huge steam engine that drives almost all atmospheric circulation, strongest in the region around the equator. The sun's radiation causes large atmospheric convective cells in the tropics, generating winds that blow equatorward as trade winds and higher-level winds that blow poleward. The trade winds and associated ocean currents were well known to early sailors who traveled from Europe and Africa to the Americas. They are still used in shipping to speed up travel and make transportation cheaper. Because of the earth's rotation, these winds do not reach the poles; the midlatitudes create the westerly winds that facilitate eastward shipping and flight. These large-scale wind systems drive most of the ocean surface currents.

The atmosphere and ocean provide many other important societal conditions that people often take for granted. Moisture evaporated from the ocean forms clouds, which carry large amounts of water that later give rise to precipitation. Almost all precipitation on earth comes from the ocean. In addition, the ocean, with its enormous heat storage capacity—thousands of times greater than the atmosphere—is the main reservoir of heat on earth.

Solar radiation and latent heat flux are not the only heat fluxes. The ocean surface emits long-wave radiation, just like the atmosphere, and to this is added the sensible heat flow that depends on temperature differences between the ocean surface and the atmosphere. The long-wave radiation emitted from the ocean surface is then reflected from the atmosphere due to greenhouse gases that, like a blanket, protect the earth from cooling down. Water vapor is the largest contributor to the natural greenhouse effect, but carbon dioxide, methane, and other gases are also important. The increasing concentration of greenhouse gases due to human burning of fossil fuels increases the reflection of long-wave radiation back from the atmosphere, which is the cause of today's global warming.

I have to tell you the ocean, a little more because there has been such a polarized discussion about long-term climate variability. It started with what was later called the "hockey stick," which showed that the global average temperature had been slowly cooling over the past two thousand years until recently, when it suddenly increased and is predicted to increase even further. Plotting the global average temperature against time shows a similar shape to a hockey stick. This graph generated a lot of discussion. Some scientists strongly argue for the existence of global warming due to human increased emissions of greenhouse gases. Others questioned this and pondered a series of scientific questions:

- How can the global mean temperature be estimated when there are so few observations from the ocean?
- Can the data and methods used be trusted?
- Given the large regional differences worldwide, are global average temperatures a good measure of climate?
- Is this a clear sign that increased greenhouse gases have changed the climate?

These questions caused many scientific disputes. It illustrated how people's beliefs can easily become polarized, giving rise to hate and group isolation. Because scientific results must be able to be reproduced by other research groups, the dispute led to largely sound results. The development showed, for example, the need for open data sources and transparent statistical methods. Although we have experienced several changes in temperature over the past two thousand years, most observations show an increasing global temperature in recent decades caused by human greenhouse gas emissions.

Will global warming protect us from a new ice age? Or will subsequent generations see a civilizational collapse due to global warming and our depletion of the earth's natural resources (Ehrlich and Ehrlich 2013)? Some argue that humanity needs to seek a new planet for future expansion because earth's resources are limited. Others warn that looking to science and technology as our salvation is dangerous (Häggström 2016), while others warn of the danger of short-term alarmist rhetoric and pessimism (Rees 2018). I now hope you, the ocean, understand that I need support. Let me put it very clearly: Will alienation from nature, carelessness, and competition cause the collapse of society? Does humanity need to export its bad behavior to other planets and leave you, the ocean, behind? How should we humans face the future?

Waiting for your guidance, I dreamed of a blue ocean where a beautiful red jellyfish swam slowly to the surface (Fig. 17.1). I felt both warmth and sadness. Jellyfish have been swimming in the sea for hundreds of millions of years. They live in the ocean during life, including bottom-anchored and free-swimming stages, and even exhibit a sleep-like state. In the free-swimming medusa stage, jellyfish sometimes have long tentacles covered with stinging cells

If I try to understand what you, the ocean, are conveying through this dream, I believe the jellyfish image symbolizes my view of a healthy sea. But I also immediately think of Medusa in Greek mythology, the monster with snakes in her hair who turned everyone who looked at her face to stone. She was originally very beautiful, but after Poseidon raped her in Athena's temple, Athena punished her by turning Medusa into a terrible woman. No wonder the jellyfish image evokes complex emotions. Is this image a reminder that people cannot see the beauty of the ocean but instead try to see you as a monster? In the myth, Perseus managed to fight Medusa, cutting off her head without looking into her eyes. Metaphorically, this myth reminds us of the strong destructive forces that exist in man. Can you, the ocean, help me understand something more about this? Your quick response, reminding me how human thinking often inverts reality and interprets things in a negative light, tells me that the beautiful red jellyfish should not be turned upside down and interpreted as Medusa. Instead, as it swims slowly in the beautiful blue ocean, it recalls the inner human resources such as intuition, dreams, and emotions that arise from our unconscious, carry new ideas, and can give a better understanding of our relationship with the marine environment.

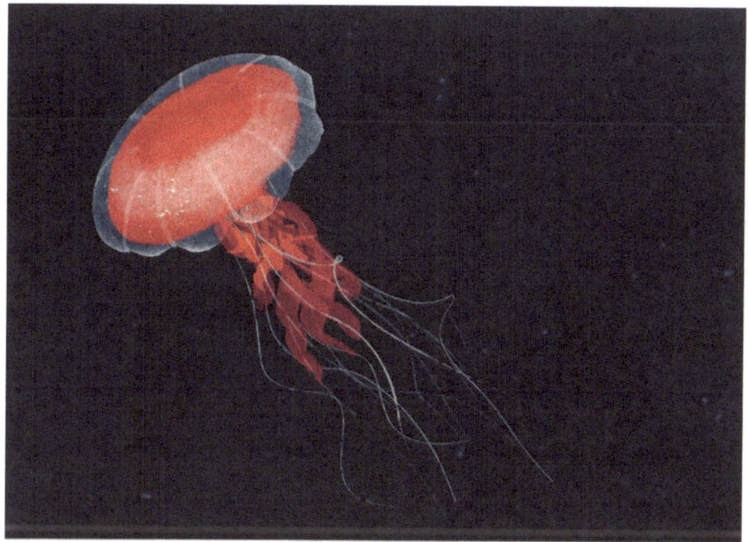

Fig. 17.1 Ocean sends an image of a jellyfish slowly rising into the sea. *Illustration* Jan Heuschele

References

Abe-Ouchi A, Saito F, Kawamura K, Raymo ME, Okuno J, Blatter H (2013) Insolation-driven 100,000-year glacial cycles and hysteresis of ice-sheet volume. Nature 500(7461):190–193

Ehrlich PR, Ehrlich AH (2013) Can a collapse of global civilization be avoided? Proc Royal Soc B Biol Sci 280(1754):20122845

Häggström O (2016) Here be dragons: science, technology and the future of humanity. Oxford University Press, Oxford, UK

IPCC (2013) In: Stocker TF, Qin D, Plattner G-K, Tignor M, Allen SK, Boschung J, Nauels A, Xia Y, Bex V, Midgley PM (eds) Climate change 2013: the physical science basis. Contribution of working group I to the fifth assessment report of the intergovernmental panel on climate change. Cambridge University Press, Cambridge, UK and New York, NY

Rees M (2018) On the future: prospects for humanity. Princeton University Press, Princeton, NJ

Chapter 18
Water Balance, Salinity, and Belonging

Abstract The water balance essential for transporting freshwater and maintaining ocean salinity is investigated here. The atmosphere transports freshwater toward the poles, and the ocean closes the cycle by transporting it back toward the equator. Most of the world's freshwater is found in the ocean. Salt comes from rocks on land and, through weathering and volcanic activity, becomes dissolved in water and transported by rivers to the ocean. The water balance and salinity are closely connected through evaporation and precipitation. Freshwater and salt are the two most important factors for life. The concerned scientist asks the ocean how modern humans can better connect to the ocean. The best way is to realize that what is going on in human bodies is also going on in the ocean. Humans and the ocean belong to each other through many important processes and from the beginnings of life itself.

Voice of the ocean: *Alienation is slowly growing like poison in your body and will disconnect you from me and life. The opposite is to belong and know that you are not alone.*

Most of the global freshwater content is in the ocean. Almost all freshwater ultimately returns to the sea through river water. While the atmosphere transports freshwater toward the poles through clouds and precipitation, the ocean transports it in the opposite direction, thus maintaining the water cycle on earth (Schmitt 1995). This is a good example of the interaction between the atmosphere and the ocean. Fresh, clean water is essential to all life, including human life. There is an increasing demand for fresh water for drinking, hygiene, food production, and industry. The availability of fresh water is critical in many arid regions and during warm periods in other places. Artificial desalination of salt water from the ocean is an important process that simultaneously produces fresh water and salt. Freshwater and salt are prerequisites for all life, and both are available from the ocean in large quantities. Desalination plants are operating worldwide, and the number of such plants is growing rapidly (Elimelech and Pillip 2011).

Evaporation, closely related to latent heat flow, is important in maintaining the water balance. As moist air rises, it cools and condenses to form clouds. The clouds

A. Omstedt, *A Philosophical View of the Ocean and Humanity*,
https://doi.org/10.1007/978-3-031-64326-2_18

and moisture are transported around the earth, returning fresh water to the surface as precipitation. Another important water cycle component is net precipitation, which is the difference between precipitation and evaporation. Net precipitation is often positive at high latitudes but negative at lower latitudes. When net precipitation is positive over land, freshwater returns to the ocean via rivers. When swimming along the shores of the Mediterranean Sea, you can sometimes feel cold water coming up from below. This groundwater from artesian springs (areas where groundwater percolates) is often colder than the surrounding seawater in the summer. Overall, net precipitation and river runoff mean that the Mediterranean Sea loses about one meter of freshwater yearly, while the Baltic Sea gains about one meter of new freshwater yearly. Therefore, the Mediterranean Sea is significantly saltier than the Baltic Sea, which has consequences for the currents in these two basins.

The water balance determines the salinity of the ocean by increasing or decreasing the transport of fresh water to the ocean. The salinity of seawater varies around an average salinity of 35 g of salt per kilogram of water. Salt has a long history and is one of the most essential minerals for the body's function and food preservation (Kurlansky 2002). It is found in large quantities in the sea and originates from rocks on land. Through weathering and volcanic activity, salts are dissolved in water and transported by rivers to the sea. Seawater takes about a thousand years to mix from the surface to the bottom, but the main chemical elements in sea salt, chlorine and sodium, have a much longer turnover time. This indicates that salt has the same chemical composition throughout the ocean. The turnover time of the salinity's chemical components is 100 million years—a long time, but still about 40 times less than the ocean's age of about 3.8 billion years. Other chemical components have different turnover times and are distributed differently (Sarmiento and Gruber 2006).

There are many ideas about the origin and development of life, but you, the ocean, were there all along. Can you say something about how modern humans can better connect with the ocean so we don't lose our connection to you?

Thank you for the new image you sent to my inner ocean at night, which looks like several rivers connected like tree branches or roots (Fig. 18.1).

This picture arouses fascination and curiosity in me. Now I see that some of the rivers depicted are red. Red can symbolize something warm or, important or emerging. Perhaps the image represents human arteries, veins, and blood? Blood transports nutrients and waste products through the body in a sophisticated manner. It transports oxygen, salt, and other chemical components around our bodies and keeps us alive and healthy. So you might be saying that humans have rivers and oceans inside their bodies and are similar to you in many ways. Do you mean that people can relate to you better by comparing what's happening in their bodies to what's happening in the ocean? And that we belong to each other ever since life began?

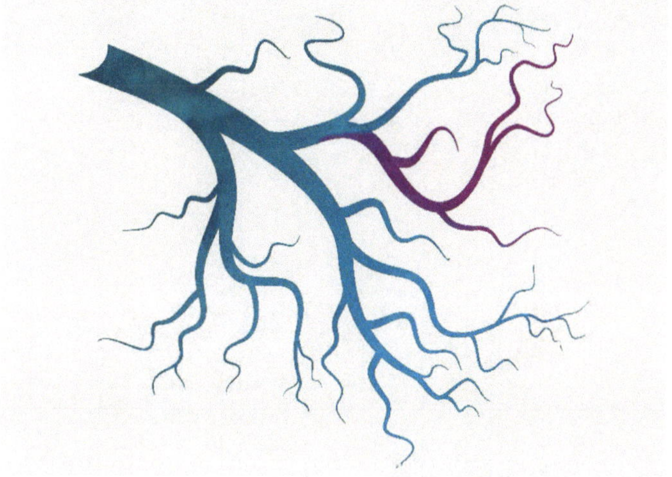

Fig. 18.1 Ocean sends an image of a network of rivers. *Illustration* Jan Heuschele

References

Elimelech M, Phillip WA (2011) The future of seawater desalination: energy, technology, and the environment. Science 333(6043):712–717

Kurlansky M (2002) Salt: a world history. Walker Publishing Company Inc., USA

Sarmiento JL, Gruber N (2006) Ocean biogeochemical dynamics. Princeton University Press, Princeton, NJ

Schmitt RW (1995) The ocean component of the global water cycle. Rev Geophys 33(S2):1395–1409

Chapter 19
Oxygen, Life, and Partnership

Abstract This chapter describes another essential chemical for life and many biogeochemical processes: oxygen. The oxygen concentrations in the ocean are declining, causing the spread of oxygen-free water. The reasons for this decline seem to be climate warming, increased nutrient transport from land, and phosphorous-leaking anoxic sediments. The question is how to protect the ocean from dying from respiratory distress. The ocean sends an image looking like a small octopus, perhaps one of the smartest animals on earth and with a different type of intelligence from that of humans. Can humans learn something new from the ocean and become more open-minded? The smart solution is to establish partnerships.

Voice of the ocean: *You think you are intelligent, but only take analytical thinking into account. Try to become wise—humanity has many complex problems, and to solve them, you need to be open-minded and not narrow-minded.*

The age of the earth is about 4.5 billion years; water began as a gas due to the extreme heat at that time. About 3.8 billion years ago, the water vapor condensed into water and filled the basins that later, through continental drift and plate tectonics, formed the ocean with the bottom topography we know today. Life began in the ocean as single-celled organisms. Animals arrived many years later, about 1 billion years ago. Through evolution, life developed, and living organisms increased in number and variety. About 600 million years ago, animal evolution split into two branches: One for humans, fish, and other vertebrates, and another for invertebrates, including mollusks like the octopus (Godfrey-Smith 2016).

Oxygen is fundamental to ocean life and many marine biogeochemical processes. Warm water cannot absorb as much oxygen as cold water, making it likely that global warming will reduce the supply of oxygen to deeper ocean layers. The ocean's oxygen content has decreased, and areas with anoxic (oxygen-free) bottom water are spreading (Breitburg et al. 2018; Diaz and Rosenberg 2008). This decreasing oxygen content is partly due to climate warming and increased eutrophication caused by transporting too much nutrients from land. Phosphore leakage from sea sediments starts in oxygen-free water, increasing eutrophication (Stigebrant and Gustafsson

A. Omstedt, *A Philosophical View of the Ocean and Humanity*,
https://doi.org/10.1007/978-3-031-64326-2_19

2007). In the ocean, oxygen is transported to deeper parts by sinking water and consumed by the decomposition of organic matter.

From your last picture, the ocean, you suggested that we could connect emotionally with you better if we compared you to what is happening in our bodies. I try to understand this and realize that the lack of oxygen must feel like the horrible feeling we get when we have severe shortness of breath. When forest fires break out during hot and dry weather on land (for example, the widespread forest fires of 2018), people must actively intervene to protect people, their houses, and animals. Such human actions are the only way to mitigate the effects of forest fires. Perhaps the same applies to the spread of oxygen-free water—people need to do something and not just stand by passively. Like in a flight emergency when one drops oxygen masks to all passengers. Could pumping oxygen-rich surface water to the deeper layers be an option?

Thanks, the ocean, for the new picture I got tonight. I had difficulty sleeping because it was such a hot summer. It took me a long time to fall asleep properly, and when I woke up, I felt frustrated and worried. The image from my dream was an octopus with many arms (Fig. 19.1).

There are many species of octopus, and fossil finds indicate that they have been in the ocean for at least 100 million years. Octopuses, from surface waters to deep layers, are often observed throughout the ocean. They breathe through their gills and skin, extracting oxygen directly from the water. They swim gracefully and are among the most intelligent animals on earth, with three hearts, nine brains, eight strong arms, and no skeleton, just a soft elastic body that can change color and shape. They have evolutionarily evolved into a separate branch from humans and developed their unique intelligence. They can spray ink to defend themselves and hide their escape but also change color and texture to hide on seabeds. They have feelings and memories that they use when solving problems.

Could a new intelligent life form develop and take over from humans, one that does not come from space but originates in the ocean? Or can humans learn to cooperate with other species and manage the ocean so that there is enough food for both octopuses and humans? I have to think more deeply about why you sent me this picture. My feelings are frustrated and troubled; the octopus can represent deep-sea mystery and subconscious resources, and its eight arms can symbolize the many possibilities of unforeseen developments. I feel stressed that you need to remind us of our inner resources and that so many options are close at hand. We don't need to plan for an alternate planet. Instead, you indicate that there is unknown intelligence to explore in the ocean and our minds, enabling new forms of partnership.

Fig. 19.1 Ocean sends an image of an octopus. *Illustration* Jan Heuschele

References

Breitburg D, Levin LA, Oschilies A, Grégoire M, Chavez FP et al (2018) Declining oxygen in the global ocean and coastal waters. Science 359(6371):46–240

Diaz RJ, Rosenberg R (2008) Spreading dead zones and consequences for marine ecosystems. Science 321(5891):926–929

Godfrey-Smith P (2016) Other minds: the octopus, the sea, and the deep origins of consciousness. Farrar, Straus, and Giroux, New York, NY

Stigebrandt A, Gustafsson BG (2007) Improvement of Baltic proper water quality using large-scale ecological engineering. AMBIO 36(2–3):280–286

Chapter 20
Plankton and Courage

Abstract After discussing oxygen, we turn to the plankton that forms the basis of the marine food web. There are many different forms of plankton with large genetic variation, giving plankton great resistance to environmental changes. Plankton provides extraordinary ecosystem services by taking up carbon dioxide, releasing oxygen, and serving as the basis of higher trophic levels. This chapter connects to the ocean through the epic poem *The Kalevala*, inspiring us to review our knowledge sources and reminding us that humans were born in the ocean. We learn that plankton is the basis of life in the ocean and that storytelling and dreaming constitute the basis of mental health.

Voice of the ocean: *You must create new global heroic poems to strengthen your mentality and support your courage. New stories are needed to reconnect me with the internal creative ocean in yourself.*

There are two main types of plankton in the ocean: phytoplankton and zooplankton. Together, they form the basis of marine life. How phytoplankton is created is an important and interesting process where chemical components such as water, carbon dioxide, and nutrients are converted into life. This creation process takes place in surface water with the help of solar radiation, as phytoplankton contain the green pigment chlorophyll, similar to land plants. All phytoplankton use photosynthesis to convert light energy into chemical energy. In this transformation, oxygen is released, and photosynthesis in plankton is one of the most important sources of oxygen in the atmosphere and ocean (Sarmento and Gruber 2006). As they grow, plankton incorporate carbon dioxide into their structures. When phytoplankton die and decompose, the process is reversed, and oxygen is used, while nutrients and carbon dioxide are returned to the water.

All organisms on earth are essentially made up of five main elements: hydrogen, carbon, nitrogen, oxygen, and phosphorus, and these elements are found in roughly constant proportions in the ocean. This simplification is used in marine ecosystem models. But it is more complicated. Silicon is found in the ocean as silicate and

A. Omstedt, *A Philosophical View of the Ocean and Humanity*,
https://doi.org/10.1007/978-3-031-64326-2_20

can explain the large presence of diatoms in many sea parts. Marine organisms also contain sulfur and, in addition, many other chemical elements in smaller quantities.

With the help of chlorophyll maps created from satellite images, it is possible to identify areas of high and low plankton activity. These areas reflect the circulation of the ocean, where productive regions are related to areas of rising water, and "desert" areas are associated with sinking water.

Cyanobacteria are a special type of phytoplankton called blue-green algae because of their color. They may have been the first oxygen-producing algae on earth. They are nitrogen-fixing and do not need phosphorus when growing. By drilling through several meters of ice and diving into the pristine Untersee Lake in Antarctica, a research team discovered that cyanobacteria could grow and generate conical stromatolites (formations up to half a meter high consisting of clay particles and organic material from cyanobacteria) on the bottom of the lake (Andersen et al. 2011). These observations indicate that cyanobacteria can survive in extreme cold and ice-covered conditions and produce oxygen despite almost no light. Through several billion years of growth of cyanobacteria, the atmosphere and water have been gradually oxygenated. Much of the life in the ocean is unknown, and new species are constantly being discovered. For example, in the 1980s, a very small plankton, Prochlorococcus, was found, which has changed our understanding of the marine food web (Chisholm et al. 1988). They are very small, related to the cyanobacteria group, and through their large availability, they are perhaps the largest primary producer of oxygen in the ocean.

Diatoms are another type of phytoplankton found globally and in many different species. Their small size means they can easily move long distances, carried by currents. One would, therefore, expect them to be well mixed throughout the ocean, but instead, they show great genetic differences even within limited areas. One reason is that they have a life cycle that includes a prolonged resting stage, which allows them to remain in sediments under severe conditions (Sundqvist et al. 2018). Their large genetic variation gives plankton resilience to environmental changes, forming a stable foundation for the marine food web.

Did you inspire me, the ocean, to study the Kalevala? Yes, you did because the poem came to me immediately when I woke up this morning (Fig. 20.1). The epic Finnish poem Kalevala originated in oral tradition but has since been written down. The first chapter describes how it all began and how the great singer Väinämöinen was born. In the beginning, Ilmatar, the goddess of air and nature, decides to go into the water, where she becomes pregnant with the ocean. After many years and complications, she gave birth to Väinämöinen, the bold bard about whom many stories are told. How to interpret this poem is described in Chap. 1 of the book.

The poem inspires us that new knowledge and insights can help thaw our hearts and create people who enjoy life, which includes sharing science, songs, and storytelling. The poem reminds us that something new was created when life was born in you, the ocean. Stories and dreams can improve mental health and inspire people to change their behavior, just as the Kalevala has brought the Finnish people together. Are you saying, the ocean, that storytelling and dreams are as important to human health as plankton is to you? I like this vision, which gives me the courage to convey your message to others.

Fig. 20.1 Ocean sends an image of how the goddess of the air, Ilmatar, becomes pregnant with the sea, the opening of the poem Kalevala. *Illustration* Jan Heuschele

References

Andersen DT, Sumner DY, Hawes I, Webster-Brown J, McKay CP (2011) Discovery of large conical stromatolites in Lake Untersee, Antarctica. Geobiology 9(3):280–293

Chisholm SW, Olson RJ, Zettler ER, Goericke R, Waterbury JB, Welschmeyer NA (1988) A novel free-living prochlorophyte abundant in the oceanic euphotic zone. Nature 334:340–343

Sarmiento JL, Gruber N (2006) Ocean biogeochemical dynamics. Princeton University Press, Princeton, NJ

Sundqvist L, Godhe A, Jonsson PR, Sefbom J (2018) The anchoring effect: long-term dormancy and genetic population structure. ISME J 12:2929–2941

Chapter 21
Ecosystems and Listening

Abstract This chapter highlights the conflict between human activity and marine ecosystems. The decline of various marine species due to human impact goes back more than a millennium but accelerated markedly in the 1950s. Hemingway wrote about the shift from sustainable to unsustainable attitudes toward fishing in his 1952 novel *The Old Man and the Sea.* Beautiful living marine resources are now mismanaged as if humanity and the ocean no longer have any relationship with each other. The ocean responds with disappointment at human lifestyles and attitudes and invites us to listen better and realize that marine resources could give much more back if managed properly.

Voice of the ocean: *I am so disappointed with humans' greedy approach to my beautiful living resources. You must listen better to me and your heart and realize my limits in supporting you. If you listen carefully to me, all my resources can be much better used and, yes, I loved the dories.*

The Northwest Atlantic, where the cold Arctic waters meet the warm waters of the Atlantic, is a biologically rich area with abundant phytoplankton, zooplankton, and krill and historically excellent fishing conditions. For many centuries, the Newfoundland banks were the site of large-scale fishing. From small open sailing and rowing boats, so-called dories, the fishermen could catch cod, haddock, and other fish species and then row to a common mother fishing vessel serving many fishermen. Not much has changed over several centuries, and the Newfoundland banks remained an area of large fish stocks. Each dory was usually handled by one or two fishermen, who filled it with fish caught with hooks and lines. Rowing and sailing these boats, about five to six meters long, required great skill, and the fishermen were well known for handling these boats in difficult and dangerous conditions, sometimes in fog, when they lost contact with the mother ship. The first documented solo sailing across the Atlantic with a five-meter-long dory was carried out by Alfred Johnson in 1876. His journey was from Gloucester, Massachusetts, USA, to Aber Castle, Wales, and lasted 58 days (Kurlansky 1999).

A. Omstedt, *A Philosophical View of the Ocean and Humanity*,
https://doi.org/10.1007/978-3-031-64326-2_21

The history of fishing and the decline of various marine species, such as sea otters, manatees, seals, whales, and cod, goes back more than a millennium (Roberts 2007). The decline of individual species often led to fishing in new areas where the marine resources could be better exploited. For four centuries, whaling was the first global industry that, after intensive commercial hunting in the late nineteenth century, made whales unprofitable to hunt. From historical studies, it is easy to trace how humans have ruthlessly altered the land and marine environment in destructive ways while maintaining an illusion of the ocean's resources as inexhaustible.

With its low fat and high protein content, cod has been the most important North Atlantic fish for centuries and could be preserved for a long time by salting and drying. Through more efficient fishing boats and catching methods, cod declined so drastically in number and size from the 1950s to 1992 that Canada was forced to cease cod fishing. The cod biomass was only one percent of previous levels at that time. The loss of the cod stocks of the Newfoundland banks was a fact. The collapse was due to massive overfishing that was only possible through technological development, environmental ignorance, and greed. The old method of fishing for cod with dories had proven to be sustainable for many hundreds of years but was outcompeted by more efficient fishing methods.

For over a century, there have been concerns about overfishing. In 1902, the International Council for the Exploration of the Sea (ICES) was formed to coordinate marine research in the North Atlantic and develop methods to estimate the size of fish stocks. During the twentieth century, most areas have been overfished (Jackson et al. 2001). This is parallel to the rapid technological development that has led to fishing vessels that can cover larger areas, the deep-freezing capacity that allows fishing fleets to stay out at sea longer, sonar and new trawling techniques that enable more efficient fishing and new ways of processing the fish, such as fish fillets or frozen fillets, making it easier to sell and cook fish. The long-term ecological change due to overfishing has caused many problems, with reduced fish stocks and reduced size distribution (Möllmann and Dieknmann 2012).

The world's supply of fish per capita is still increasing due to the rapid growth of aquaculture, which, since 2014, has accounted for almost half of all fish for consumption. The world's wild fish stocks have not improved overall, and about a third of commercial fish are fished at unsustainable levels (FAO 2016). Added to this is extensive illegal fishing, making estimating fish stocks difficult. As much as four-fifths of the world's catch cannot be assessed, and illegal, unreported, and unregulated fishing is a growing problem. Illegal fishing includes catching sharks that are thrown overboard after finning them, as well as by-catches that are thrown overboard because they are not commercially viable. Baltic cod stock has declined for many reasons, including overfishing, inappropriate fishing methods, and eutrophication. Today, this fish population seems to have collapsed, and only small Baltic cod remain, while industrial fishing is emptying the Baltic herring. Another example of stock decline is the collapse of the North Sea herring in the latter part of the twentieth century. This example raises questions about managing an individual stock and its relationship to the ecosystem (Dickey-Collas et al. 2010). Again, this is a clear example of human's careless relationship with the ocean.

The future of the ocean is closely related to the future of fish stocks, and there is an urgent need to rethink our fisheries. The world's fish stocks have never been exploited as intensively as in the last 50 years (World Ocean Review 2 2013). Non-industrial fishing is conducted from small vessels, often in developing countries. Industrialized fishing methods can instantly process, package, and deep-freeze large quantities of catch, exploiting global fish stocks far beyond sustainable levels. Fishing has also gone deeper into the sea, and trawl nets are now used at depths as great as 2,000 m, with the risk of severe effects on, for example, coral reefs and bottom-dwelling organisms.

In *The Old Man and the Sea* by Ernest Hemingway, the shift from sustainable to unsustainable fishing had already been imagined in 1952, when the first edition was published (Hemingway, 1952):

> *He [i.e., the older man] always thought of the sea as la mar, which is what people call her in Spanish when they love her. Sometimes, those who love her say bad things about her, but they are always expressed as though she were a woman. Some of the younger fishermen, those who used buoys as floats for their lines and had motorboats, bought when the shark livers had brought much money, spoke of her as el mar, which is masculine. They spoke of her as a contestant, place, or enemy.*

In science, sudden and rapid changes in marine ecosystems, so-called regime shifts, have been increasingly reported, illustrating the need for a longer perspective when evaluating overfishing. Analyses of bottom trawl catches landed in England and Wales from 1889 showed a ninety-four percent reduction (Thurstan et al. 2010). Bottom-dwelling fish species and the bottom ecosystem have undergone extensive changes since the nineteenth century. This can be illustrated by comparing observations made by Carl Georg Johannes Petersen (1860–1928) during 1884–1886 in the eastern Kattegatt with more recent data. Here, one can see big dramatic changes, mainly caused by bottom trawling (Josefson et al. 2018).

What are your emotions, the ocean, when you see how people have mismanaged your resources and foundations in such a disappointing way? It is as if humanity and the sea no longer relate. Why are people so stupid? Beautiful marine ecosystems and delicious fish should and can be harvested sustainably. This clearly illustrates the need to change our thinking about marine resource management. I liked old Santiago's view of the sea and preferred the time when small boats fished selectively, but now fishing is industrial. Perhaps I have forgotten that the earth's population has grown and needs more food than before. What do you say, the ocean?

You seem to be quiet. My head is noisy, with many voices from everything I've talked and read about. When I stop thinking so much, my intuition wakes up, and I can connect with you more easily. To be more receptive, I need to become better at listening. Listening is hard when my head is full of competing voices and ideas. Maybe I need to concentrate more on our dialogue (Fig. 21.1). Listening requires me to pay more attention to the human-ocean relationship. I can't read every interesting new book or watch every exciting new movie, so I must improve my ability to choose what to pay attention to. As an academic, I am trained to research and argue for my ideas, but now I have to listen and ask open-ended questions. It also means that I need to understand my inner self better. I have always had a strong will to live and participate in the world. Maybe this is what it's all about for you too? I need to learn more about listening to discern your opportunities and realize that marine resources can provide much more if managed properly.

Fig. 21.1 Ocean sends an image of its wealth. *Illustration* by Jan Heuschele

References

Dickey-Collas M, Nash RDM, Brunel T, van Damme CJG, Marshall CT, Payne MR, Simmonds EJ (2010) Lessons learned from stock collapse and recovery of North Sea herring: a review. ICES J Mar Sci 67(9):1875–1886

FAO (2016) The state of world fisheries and aquaculture 2016: contributing to food security and nutrition for all. FAO Fisheries and Aquaculture Department, Food and Agriculture Organization of the United Nations, Rome, Rome, Italy

Hemingway E (1952) The old man and the sea. Jonathan Cape, London, UK

Jackson JB, Kirby MX, Berger WH, Bjorndal KA, Botsford LW, Bourque BJ, Warner RR (2001) Historical overfishing and the recent collapse of coastal ecosystems. Science 293(5530):629–638

Josefson AB, Loo L-O, Blomqvist M, Rolandsson J (2018) Substantial changes in the depth distributions of benthic invertebrates in the eastern Kattegat since the 1880s. Ecol Evol 8(18):9426–9438

Kurlansky M (1999) Cod. A biography of the fish that changed the world (in Swedish). Ordfront's publisher. Stockholm

Möllmann C, Diekmann R (2012) Marine ecosystem regime shifts induced by climate and overfishing: a review for the northern hemisphere. In: Woodward G, Jacob U, O'Gorman EJ (eds) Advances in ecological research, vol 47. Academic Press, London, UK, pp 303–347

Roberts C (2007) The unnatural history of the sea. Shearwater Books-Island Press, Washington, DC

Thurstan RH, Brockington S, Roberts CM (2010) The effects of 118 years of industrial fishing on UK bottom trawl fisheries. Nat Commun 1:15

World Ocean Review 2 (2013) Living with the ocean. The future of fish: the fisheries of the future. Maribus gGmbH in cooperation with Future Earth, Kiel Marine Sciences, Hamburg. Available at https://worldoceanreview.com/en/

Chapter 22
Non-living Ocean Resources and Hope

Abstract After studying humans' effects on living marine resources, we now focus on non-living marine resources. Mineral oil formed from phytoplankton illustrates the close connection between living and non-living resources and represents today's most important fossil energy source. The search for new oil and gas sources is intense, and increasing interest in ocean exploration for oil and gas, even in deep ocean areas, entails many environmental risks. Will humans' hunger for oil and gas destroy the ocean or be satisfied in an environmentally wise and sustainable manner? The ocean reminds us that there is light and shadow deep inside all of us and that we can easily give in to destructive desires. A mental change is needed, encouraging humans to work together to generate global cooperation and hope.

Voice of the ocean: *Find the gold inside yourself and give it away to improve your mental health. Self-absorption and greed will turn you into monsters. Working together with all will give hope.*

Ocean water is a unique mixture of almost all chemical elements supplied by rivers, originating from the atmosphere and volcanic activity and in concentrations that vary over a wide range. The non-living marine resources include salt, sand, gravel, phosphate, diamonds, gold, manganese, copper, nickel, iron and cobalt, crude oil, and gas. Both oil prospecting and the mining of seabed minerals are lucrative and entail great risks.

The global carbon cycle, with organic and inorganic parts, plays a fundamental role in the earth's system. If the carbon cycle were in balance, carbon dioxide in the atmosphere and pH in the ocean would not vary over several years. Instead, it is out of balance due to man's burning of fossil fuels and land use (Gattuso and Hansson 2011). Fossil fuels generated by biological life over hundreds of millions of years are now used very quickly. Today, fossil carbon dioxide emissions are the largest part of the global carbon dioxide turnover. Of these emissions, about half stay in the atmosphere, and the rest is taken up by plankton in the ocean and plants on land.

Human's direct impact on the earth's system can be illustrated by two important observations that both show that the system is out of balance:

A. Omstedt, *A Philosophical View of the Ocean and Humanity*,
https://doi.org/10.1007/978-3-031-64326-2_22

- The carbon dioxide content in the earth's atmosphere increases.
- The pH of the ocean's surface water is falling.

At the beginning of the industrial era, around 1750, the atmospheric carbon dioxide content was about 277 parts per million (ppm). In 2015, the level passed 400 ppm and has since increased (Le Quéré et al. 2018). The ocean's absorption of the carbon dioxide produced by humans is an important service at the cost of its acidification. Long-term ocean acidification measurements are lacking, and many shorter measurements show great variation in time and space. However, a few measuring stations with observations over a few decades show decreasing pH values of about 0.1 pH units from values around 8.1. The ocean is thus still basic, but the pH is falling faster in various places and is expected to continue to fall. An example of faster acidification is the Arctic, where the pH drops three to four times more quickly, as melting sea ice promotes rapid absorption of atmospheric carbon dioxide (Qi et al. 2020).

The key to understanding the ocean's carbon system is knowing its chemical reactions in water (Schneider and Mueller 2018). This chemistry describes how dissolved inorganic carbon, which consists of three components (carbon dioxide, bicarbonate, and carbonate, the sum of which is defined as the total dissolved inorganic carbon), reacts in water. In addition, total alkalinity is added to the ocean's carbon system, defined as the excess of substances that absorb hydrogen ions over those that emit hydrogen ions. The total alkalinity measures the seawater's ability to counteract changes in the acid–base balance. The main process controlling the cycle of total alkalinity is the biological production/dissolution of calcium carbonates. Reduction or production of calcium carbonate occurs in the surface water when phytoplankton with shells (such as diatoms) grow or become mineralized as they sink to deeper layers and sediments. At the ocean surface, total alkalinity changes primarily due to evaporation and precipitation, as well as changes in salinity, allowing for a close relationship between total alkalinity and salinity. This is not the case in coastal seas such as the Baltic Sea, where the total alkalinity is influenced by the rivers in the southern parts, which are rich in calcium carbonate. In contrast, the rivers in the northern Baltic Sea are poor in calcium carbonate (Hjalmarsson et al. 2008).

Mineral oil is mainly formed from large amounts of phytoplankton that have sunk to the seabed as dead material, while coal and natural gas are primarily from land plants (World Ocean Review 3 2014). The dead phytoplankton accumulated on the ocean floor and mixed with sand and mud, becoming sludge that slowly, under high pressure, is converted into mudstone and then, under high temperatures, into oil. Mineral oil is a mixture of many different chemical components used after processing in refineries in products such as petrol, diesel, and plastics. Global energy consumption is increasing, and the International Energy Agency (IAE) estimates that energy consumption will continue to grow in the coming decades. Mineral oil is the most important fossil energy source, followed by coal and natural gas. Together, they are today the dominant sources of global energy consumption. Alternative energy sources such as hydropower, nuclear power, and renewable energy comprise a significantly smaller, albeit growing, part of the global energy supply. The world community is thus heavily dependent on fossil fuels formed over hundreds of millions of years.

The search for new sources of oil and gas is intense, with an increased interest in the ocean. Offshore drilling for oil and gas has been done for over a century, starting in shallow shelfs and continuing to deep (over 400 m) and ultra-deep (over 1500 m) ocean basins. Today, offshore production accounts for approximately one-third of global oil and gas production. With the help of new technology, it is now possible to discover oil and gas at much greater depths, and new potential oil and natural gas fields have also been found at such depths. Extraction of fossil energy at great depths and in the Arctic will entail enormous environmental risks. Therefore, long-term strategies for marine conservation, clean energy, and sustainable use of marine resources are needed.

Now I must ask you, the ocean, if human's hunger to explore and exploit the ocean will ultimately destroy us all? I know scientists have presented many examples of overcoming the "tragedy of the commons" by working together, but how can this be done for the good of the ocean? Yesterday when I went to bed I couldn't sleep. My head and body were full of energy, just waiting for guidance. Several hours after I finally fell asleep, I awoke with an image of a deep valley with something in it that glittered like gold (Fig. 22.1).

As I reflected on this image, I was surprised and alert. The gold reminded me of the ring from the book *The Lord of the Rings* by John Ronald Reuel Tolkien (1892–1973), a story about the struggle between good and evil. When Tolkien wrote this famous novel (Tolkien, 1954a, b; 1955), he was influenced by the effects of industrialization and world war. My feelings are awakened for the brave Frodo and the unhappy Gollum in this story. Frodo was given the ring by his uncle Bilbo. The wizard Gandalf tells Frodo about the evil Lord Sauron, who needs the Ring to access his full destructive power. Gandalf advises Frodo to destroy the ring. Frodo embarks on a long and dangerous journey to destroy the ring. But the ring's power is great, and in the end, Frodo cannot resist the ring's power and puts it on his finger. At that moment, Gollum, the ring's former owner, appears, and in a fight, he bites off Frodo's finger with the ring. Gollum falls into the doomsday chasm with violent fire and destroys the ring and, with it, the power of evil.

Perhaps you, the ocean, are trying to tell me that deep within all of us, there is a light part, represented by Frodo, but also a shadow part, defined by Gollum. Both parts can give in to destructive desires for power and gold or seek only their satisfaction. Instead, looking at this as a metaphor, *The Lord of the Rings* illustrates how the greedy nature in our minds needs to be destroyed to save the ocean. The ocean, is that what you're trying to convey?

Fig. 22.1 Ocean sends an image that points to hidden resources. *Illustration* Jan Heuschele

References

Gattuso J-P, Hansson L (2011) Ocean acidification. Oxford University Press, Oxford, UK

Hjalmarsson S, Wesslander K, Anderson LG, Omstedt A, Perttilä M, Mintrop L (2008) Distribution, long-term development and mass balance calculation of total alkalinity in the Baltic Sea. Cont Shelf Res 28(4–5):593–601. https://doi.org/10.1016/j.csr.2007.11.010

Le Quéré C, Andrew RM, Friedlingstein P, Sitch S, Hauck J et al (2018) Global carbon budget 2018. Earth Syst Sci Data 10(4):2141–2194

Qi et al (2020) Climate change drives rapid decadal acidification in the Arctic Ocean from 1994 to 2020. Science 377:6614

Schneider B, Mueller JD (2018) Biogeochemical transformations in the Baltic Sea. In: Observations through carbon dioxide glasses. Springer Oceanography. Springer International Publisher AG. https://doi.org/10.1007/978-3-319-61699-5

Tolkien JRR (1954a) The lord of the rings: the fellowship of the ring. George Allen & Un, London, UK

Tolkien JRR (1954b) The lord of the rings: the two towers. George Allen & Unwin, London, UK

Tolkien JRR (1955) The lord of the rings: the return of the king. George Allen & Unwin, London, UK

World Ocean Review 3 (2014) Living with the ocean. Marine resources: Opportunities and risks. Maribus gGmbH in cooperation with Future Earth, Kiel Marine Sciences, Hamburg, Germany. Available at https://worldoceanreview.com/en/

References

Chapter 23
Human Interaction and Vision

Abstract The previous chapter treated the need to work together; now, we look at the growth in population and change in living conditions over past centuries. Before 1800, the global population was less than a billion, whereas today it is about 8 billion and still growing. Coastal regions have been extensively explored and are becoming home to large and growing populations. Will global civilization collapse as people work disparately in various directions without addressing how to achieve healthy and sustainable coordinated ocean management? The ocean reminds us that we need a shared vision to improve living conditions for all. The UN Sustainable Development Goals for 2030 articulate a vision that could help us develop a much better relationship with the ocean.

Voice of the ocean: *Anything could happen in the future, and without a vision, you may lose what is most important in life. Everything is dependent on how you interact. Your great challenge is to support one another and nature. If you do so, your carrying capacity will grow with benefits for all of us.*

The services the ocean offers humanity are extensive and cannot be measured in money. The ocean is the foundation of all life on earth and supports it in countless ways. The ecosystem service concept is misleading because it only points in one direction and does not capture the ocean-human connection. It is better to talk about ecosystem functions, and we should think about our role vis-à-vis the ocean. The economic value of the ocean is also difficult to estimate, and how it should be calculated is still disputed (World Ocean Review 4 2015). Estimating, for example, the costs of ocean acidification and evaluating its effect on fisheries, marine ecosystems, coral reefs, etc., is today impossible due to the many complex processes that need to be included in such a calculation. Human exploitation of the ocean is a sad story of unsustainable practices. Most of us know better as individuals, but collectively, we are careless and persist in viewing the ocean as an infinite resource that can be exploited without acknowledging its limitations.

Human exploitation of the ocean is as long as human history. Most people think of the sea as blue, endless, and dangerous, but it is a place of hard work and great challenges for sailors. The ocean has historically been the natural link between people

A. Omstedt, *A Philosophical View of the Ocean and Humanity*,
https://doi.org/10.1007/978-3-031-64326-2_23

from different countries. Many people worldwide, in history and today, make their incomes and livelihoods along the world's coasts. Poets, writers, and painters have filled our imaginations with marine images of wonder, glory, and horror. This is evident in many books, outstanding in Greek mythology and literature, such as *The Odyssey* by Homer. Or from Botticelli's beautiful painting of the birth of Venus in the sea.

In 1895, Captain Joshua Slocum embarked on a solo sailing voyage around the world aboard a rebuilt 11-m oyster sloop, *Spray*. In 1900, he published *Sailing Alone Around the World*, inspiring many sailors and adventure seekers (Slocum, 1900/2017). Slocum started by rebuilding *Spray* over 13 months in 1893 and 1894. From 1885 to 1898, he sailed across the Atlantic twice, passed Cape Horn through the Strait of Magellan, and crossed the Pacific Ocean. He visited Australia and South Africa and crossed the Atlantic Ocean once more before returning to Massachusetts as the first person to sail around the world alone. Slocum's book describes many sights and situations typical of sailing on the ocean, such as blue waters, fog, storms, doldrums, the danger of running aground, solitude, navigation problems, and pirates. His journey demonstrated that, in the late nineteenth century, it was possible to travel around the world single-handedly. The ocean could no longer be regarded as infinitely large.

Over the past two centuries, maritime transport has evolved from sailing ships, fishing vessels, and smaller boats to steamships powered by coal and wood to ships and fishing vessels powered by oil and natural gas. At the same time, shipping capacity increased exponentially (Stopford 2009), and trade routes around the globe were shortened by the construction of canals, for example, Suez (1869), Panama (1914), and Kiel (widened 1907–1914). The rapid industrialization after World War II increased fishing with the development of fishing methods on an industrial scale and through fierce competition. Many nations industrialized their fisheries, replacing small-scale fishing boats with fewer but larger, government-subsidized vessels (Finley 2016). Parallel to the history of shipping, piracy of various kinds has occurred, and jurists have regarded pirates as the enemy of all (Heller-Roazen 2009).

At the beginning of the nineteenth century, the world's population was around one billion. Population growth was slow until 1900, but by 1950, the global population had reached 2.5 billion. After 1950, population growth increased dramatically, and today, eight billion people live on earth, mostly in Asia (published online by Our World in Data; www.ourworldindata.org). At the end of the twentieth century, the world's population is expected to reach around 11 billion. The rapid population growth is strongly associated with reduced mortality. The proportion of the world's population dying before the age of five has fallen dramatically and is one of the factors why world life expectancy has more than doubled since 1900. The number of births per woman has fallen from about six births per woman in 1800 to about 2.5 births per woman today; in the future, births are expected to decrease further.

The physician Hans Rosling (1948–2017) emphasized the importance of using up-to-date statistics and having two thoughts in mind simultaneously: current and past trends (Rosling et al. 2018). The world's population has increased significantly since the nineteenth century due to improvements in healthcare, water and food supply, child care, and education. Over the past 20 years, the number of people living in extreme poverty has been halved. That eight billion people today live in much better conditions than far fewer people did in the past is an example of great progress. An important conclusion from population statistics studies is that social development that gives poor people better incomes and education is the best way to reduce population growth.

The majority of the world's population today lives in cities. The United Nations estimates that in 2018, 55% of the world's population lived in cities and surrounding areas. By 2050, as many as 68% are projected to live in cities and their suburbs. By comparison, in 1950, only 30% of the world's population was urban. Many of the world's major cities are on the coasts, and coastal regions are home to growing populations. Tokyo is the world's largest city, with 37 million inhabitants, followed by Delhi, Shanghai, Mexico City, and Sao Paulo. Urban and rural living conditions differ greatly, with the risk of greater inequality in living conditions, education, and access to social services.

It's morning, and I'm reflecting on whether or not global civilization will collapse. I woke up in sadness and remembered our dialogue, the ocean. As a scientist, I can name many world threats that could lead to collapse. My mind is filled with so many warning and fearful voices. Scientists and society are working in different directions without addressing how to develop healthy, sustainable, and coordinated ocean management. This reminds me of the biblical story of the *Tower of Babel*. It tells about how all the world's people spoke the same language. They decided to work together and build a tower to reach the sky. God intervened and changed this so that everyone began to speak different languages. The people could no longer understand each other to cooperate, and the *Tower of Babel* could not be completed. Will our extreme specialization result in a similar disintegration, or will it lead to new solutions to today's problems?

I've been thinking a lot about our last contact and trying to keep quiet and listen to my inner ocean. But I must tell you something more. It is clear that humans have made enormous progress in the last 200 years, but the price of this progress has also been high. Human population growth can be seen as a sign of increased prosperity and human creativity. But there are also major destructive changes in the world that many are worried about and trying to address. The most ambitious vision is the UN's 17 goals for sustainable development. They represent conditions for a better life, including human rights and dignity, for all people (the goals and their opposites are discussed in Chap. 27). The vision and goals must be understood as a whole, not individually, and are truly worth striving for. So many goals and such an important mission require a completely new way of thinking—can you, the ocean, suggest something that might be missing in these goals?

The ocean, you are silent, but now you send me a completely unexpected image (Fig. 23.1). It looks like a boat anchored to an island, tied to it with a rope. Can you say anything more? No, you are quiet and may be waiting for my reaction. Reflecting on the image, I love the boat and am delighted at its integration with nature. If this image is meant for us, for humanity and you, the boat can be a metaphor for a better life, floating safely on the sea. Perhaps you suggest adding trust and pride to the UN's vision, which, like an ark, can carry us to a safer future. The picture makes me happy and gives me hope that people can develop a better relationship with the ocean.

Fig. 23.1 Ocean sends an image that shows the beauty of a harmonious relationship. *Illustration* Jan Heuschele

References

Finley C (2016) The industrialization of commercial fishing, 1930–2016. In: Oxford Research Encyclopedia of environmental science. Retrieved from https://oxfordre.com/environmentalscience/view/, https://doi.org/10.1093/acrefore/9780199389414.001.0001/acrefore-9780199389414-e-31

Heller-Roazen D (2009) The enemy of all: piracy and the law of nations. Zone Books, Brooklyn, NY

Rosling H, Rosling Rönnlund A, Rosling O (2018) Factfulness: ten tricks to help you understand the world. Nat Cult, Stockholm (in Swedish)

Slocum J (1900/2017) Sailing alone around the world. Dover Publications, New York, NY. Dover edition originally published 1956; reprinted 2017

Stopford M (2009) Maritime economics. Taylor and Francis, Hoboken, NJ

World Ocean Review 4 (2015) Sustainable use of our oceans: making ideas work. Maribus gGmbH in cooperation with Future Earth, Kiel Marine Sciences, Hamburg, Germany. Available at https://worldoceanreview.com/en/

Chapter 24
Climate Change and Harmony

Abstract The previous chapter discussed the unprecedented rise in population growth over the past century. This chapter addresses anthropogenic climate change due to increased GHG emissions caused by burning fossil fuels—among the drawbacks of industrialization. How carbon dioxide, one of the GHGs, could influence air temperature was already understood in the late nineteenth century. It was not until the late 1980s and the first assessment report of the Intergovernmental Panel on Climate Change (IPCC), published in 1990, that concern about global warming spread to a larger group of people. The ocean is under pressure from many stressors, and restoring healthy marine environments is needed to meet these threats. The image of a healthy ocean suggests a likeness between the ocean and humans, for whom health can be physical and mental. The ocean challenges us to build such health and feel the ocean's joy, which can create greater harmony between humans and the ocean.

Voice of the ocean: *A good life is beautiful, and compassion gives a great deal back. I challenge you to build physical and mental harmony between us.*

Climate has worried people for many generations (Lamb 1995). The changes have been seen as unexpected, with rapid variations causing difficult living conditions through, for example, severe winters, crop failure, unexpected frost, ice, icing on ships, floods, droughts, and storm damage on land and sea. During the nineteenth century, it was realized that carbon dioxide absorbed heat radiation (Eunice Foote in 1856 and three years later John Tyndall see Jackson 2019; Arrhenius 1896). Variations in atmospheric carbon dioxide content were also discussed as a possible explanation for major climate variations, such as ice ages. The subject was then forgotten, and it was mainly believed that human influence was insignificant compared to natural variations through solar activity and ocean circulation (Maslin 2004). In the late 1950s, the chemist Charles David Keeling (1928–2005) began direct measurements of atmospheric carbon dioxide, and soon these measurements were clear signals of increased levels of carbon dioxide in the atmosphere. Several scientists predicted that this was due to burning oil, coal, and gas and would lead to global warming.

A. Omstedt, *A Philosophical View of the Ocean and Humanity*,
https://doi.org/10.1007/978-3-031-64326-2_24

It was not until the 1980s that new estimates of global average temperatures showed that global warming was occurring. It was the IPCC that came to show that humans have caused global warming through emissions of greenhouse gases (IPCC 2023). Its effects on a regional level were evaluated through various regional initiatives in, for example, the Arctic and the Baltic Sea area (BACC II Authors Team 2015). Climate research has shown that freely available observations, data products, models, and regular international scientific assessments are fundamental for credible communication between researchers and society. Here, the IPCC's working methods have been successful, and the message about human impact on the climate is now well known in society. Not as well known is that fewer and fewer people are dying from climate-related natural disasters. Over the past hundred years, annual climate-related deaths have declined sharply, which is a major success for the adaptability of modern society.

With increasing exploration, the marine environment is exposed to multiple stresses from many actors, and climate change caused by increased greenhouse gas emissions is a factor that interacts with other threats. There is growing concern that the marine system's resilience will decline if management fails to restore a healthy marine environment. The ocean and human health are inextricably linked. But what does a healthy ocean mean? Healthy is used metaphorically in this context, and metaphors are the language of art and dreams. Here, the concept of a healthy ocean suggests a similarity between the ocean and humans, for whom health can be physical, mental, or even economic. Today, no global observing system is available to assess ocean health, and there are ongoing discussions in science and management about how to measure it.

Dear ocean, as you indicated, people could relate to you better if they reflected on what is happening in their bodies and souls. But since measuring your physical and mental health is difficult, I ask you what the most important factors are in understanding your health. I guess the important human factors are weight, body temperature, blood pressure, and emotions, but what would be the best indicators to measure your health?

This morning, I got a new inner image of a seabird flying just above the sea's surface. The picture evokes fascination, happiness, and excitement (Fig. 24.1). The seabird flies so beautifully, close to the surface, and takes advantage of winds right over the waves. It flies with great precision and in complete harmony with the ocean surface. Sailors once considered a seabird flying over the water the navigator's best friend, as these birds were mostly seen near coasts and signaled that land was near. The bird in the picture you sent me, the ocean, looks like an albatross, a species believed to harbor the souls of lost sailors. Killing them was thought to bring bad luck. Here, you have sent a picture of an albatross in complete harmony with the ocean—something completely different from the picture of the seabird that died from plastic waste. Perhaps we should understand harmony as we understand harmony in music (Snieder and Schneider 2016), where collaboration between voices can create something greater. Do you think harmony between people and the ocean should be the best indicator of ocean health? The opposite—destructive human environmental behavior—leads to human isolation and death, as described so well in Samuel Taylor Coleridge's 1798 poem, *The Rime of the Ancient Mariner*.

Fig. 24.1 Ocean sends an image of an albatross flying in full harmony. *Illustration* Jan Heuschele

References

Arrhenius S (1896) On the influence of carbonic acid in the air upon the temperature of the ground. Philosophical Magazine and Journal of Science, Series 5(41):237–276

BACC II Author Team. (2015). Springer regional climate studies. Second assessment of climate change for the Baltic Sea basin. Springer International Publishing, Cham, Schweiz

IPCC (2023). Summary for Policymakers. In: Climate Change 2023: Synthesis Report. Contribution of Working Groups I, II and III to the Sixth Assessment Report of the Intergovernmental Panel on Climate Change [Core Writing Team, H. Lee and J. Romero (eds.)]. IPCC, Geneva, Switzerland, pp. 1–34, https://doi.org/10.59327/IPCC/AR6-9789291691647.001

Jackson, R., (2019). Eunice Foote, John Tyndall and a question of priority. Notes and Records. The Royal Society Journal of History of Science. https://doi.org/10.1098/rsnr.2018.0066

Lamb, H. H. (1995). Climate, history and the modern world (2:a uppl.). Routledge, New York, NY.

Maslin M (2004) Global warming: A very short introduction. Oxford University Press, Oxford, UK

Snieder R, Schneider J (2016) The joy of science: Seven principles for scientists seeking happiness, harmony, and success. Cambridge University Press, Cambridge, UK

Chapter 25
Scenarios, the Future, and Simplicity

Abstract Many are trying to foresee what will happen in the future. Numerical calculations of societal changes often rely on statistical methods that cannot predict the future. Many try to relate various parameters to one another, hoping to gain knowledge of what is to come. Weather forecasting can provide information about conditions some days in advance. Carbon dioxide emissions and other anthropogenic GHGs must be reduced to prevent global warming. However, the global earth system models applied in climate change assessments rely on prescribed emission storylines, and their projections are conditional on the assumed forcing. A major question for society is how to develop a greener and more sustainable future. The ocean responds by sending an image of barnacles adhering to a cliff, indicating that humans need to slow down, learn more from the ocean, and build durable and trusting relationships.

Voice of the ocean: *When you feel like something is chasing you, remember that the best thing is to slow down, turn around, and reflect on what is happening. Time is not running out; rather, time is bringing new challenges and opportunities for reflection.*

As already mentioned, many factors threaten the well-being of the ocean, threats caused by

- Climate change,
- warming,
- ocean acidification,
- overfertilization,
- pollution,
- plastic waste,
- ammunition dumping,
- overfishing and destructive fishing methods,
- constructions along the coasts,
- water and nuclear power plants,
- wind farms,
- bridges and tunnels.

A. Omstedt, *A Philosophical View of the Ocean and Humanity*,
https://doi.org/10.1007/978-3-031-64326-2_25

When examining ocean changes, one should expect that the marine system, in general, is affected by many factors simultaneously. The considered time scale is also important—if you want to say something about the next hundred years, for example, it is a good idea to adopt a historical perspective. A time scale of a thousand years illustrates, for example, how the climate has varied over different centuries.

But just knowing the past and present is not enough to determine what might happen in the future. The current situation and past trends do not necessarily indicate what may happen. Of course, historical knowledge and data from past times are important for understanding recent developments in nature and society. Still, trends can only be used when applied to the period from which they originated. Outside the observation period, a trend cannot be used to make predictions. Then, a deeper understanding of how nature and society work together is needed. For example, if one studies temperature observations from January to August, the observed trend will give a completely wrong prediction for the temperature in the coming winter. During the winter, the temperatures decrease, which is well known and can be called a system understanding. Statistical methods and probability analyses are used in many different applications in society, and there is a general belief that many predictions can be made more accurately if you count on average values. This is sometimes called the "wisdom of the crowd," but there are always surprises—for example, most people who invest money in stocks cannot predict market crashes. Without a deep understanding of systems, predictions can be very misleading.

Another way to tell something about the future is to use statistical relationships. It is often possible to find connections between something you are interested in and other variables you can observe. It is easy to correlate multiple datasets with each other and get incorrect and false results. In northern Europe, for example, winter temperatures are statistically linked to the large-scale atmospheric circulation. However, this correlation changes over time and is therefore unsuitable for predictions (Omstedt and Chen 2001). Most predictions about future developments based solely on statistical methods are consequently uncertain.

Many different atmospheric and ocean patterns are stochastic, that is, random. The situation is different with astronomical predictions, which can provide reliable information many years ahead, such as how the moon's orbit around the earth will vary. Changes in stochastic flows cannot be predicted more than a short time ahead—after about a week, weather forecasts provide no information. Because climate models are based on weather models, they cannot be expected to have much predictive value, although they can calculate the statistical properties of the weather. A major problem in the experiments with the climate models is that we do not know the details of the assumed forcing, i.e., how the earth and its population will develop. Climate models are therefore not used as forecast models but rather to investigate the climate's sensitivity to certain prescribed emission amounts. Work on estimating the earth's climate sensitivity has been developed successfully, and today, many calculations are freely available for various emission scenarios. Reducing this uncertainty is a major research challenge; meanwhile, ensemble means and expert judgments are used to say something about the future (Meier et al. 2018).

The various calculations or scenarios aim to capture a wide range of possible changes in future emissions, ranging from reductions to substantial increases (O'Neill et al. 2017). The calculations illustrate extensive changes in the climate with major consequences for society. But there is no easy way to predict the future in complex systems. How we manage and shape societal development depends on what we do, and the models can only indicate the consequences of various prescribed emissions. Will society strive for a greener, more sustainable development path with reduced carbon dioxide emissions, leading to a healthier environment? Or will future development lead to a fragmented world of regional rivalry and continued high consumption of energy and natural resources? I need inspiration and ask if you, the ocean, can guide me.

Thanks for this interesting picture—it looks like barnacles attached to a rock (Fig. 25.1). This reminds me of when I was young and swam on the Swedish west coast. If I was careless, I could easily injure the feet of barnacles and start bleeding. In different phases of the barnacle's life cycle, these small ocean creatures transform from being free-swimming larvae to attaching to all sorts of things, such as rocks, clams, whales, boats, and buoys. So what are you trying to tell me? The image evokes feelings of simplicity and caution. Barnacles have lived for millions of years and adapted to rapid changes in water levels and harsh ocean conditions. They are a problem for humans because they attach to boats and anything that ends up in the ocean, but this is not what you are trying to tell me. Instead, I have to think about how I can interpret the image metaphorically. The barnacles begin life as free-swimming larvae. They settle down and build protective limestone houses to filter the water when they mature. Perhaps you mean people should slow down, strive for simplicity, and build sustainable and trusting relationships? Yes, people have a lot to learn from you, the ocean.

Fig. 25.1 Ocean sends an image of barnacles on a rock. *Illustration* Jan Heuschele

References

Meier HEM, Edman MK, Eilola K, Placke M, Neumann T et al (2018) Assessment of eutrophication abatement scenarios for the Baltic Sea by multi-model ensemble simulations. Front Mar Sci 5(28), Article 440

O'Neill BC, Kriegler E, Ebi KL, Kemp-Benedict E, Riahie K, Rothman DS, Solecki W (2017) The roads ahead: narratives for shared socioeconomic pathways describing world futures in the 21st century. Glob Environ Chang 42:169–180

Omstedt A, Chen D (2001) Influence of atmospheric circulation on the maximum ice extent in the Baltic Sea. J Geophys Res 106(C3):4493–4500

Chapter 26
Summary, Part II

Abstract The book's second part extends the knowledge from the first part by connecting marine science to the arts through communication between a marine scientist and the ocean. In this summary, the earlier chapters are reviewed. In the dialog, we are reminded that the most meaningful part of life is nurturing new life under healthy living conditions. Developing healthy and joyful living conditions is possible, and with the growing human population, it should be humanity's primary task. Destructive thinking and behavior scare us and will not promote the future development of a more sustainable lifestyle. Human development and culture brought about by art, literature, philosophy, and science have extraordinary capacities and can provide the impetus for necessary mental change.

The book's second part describes the threats the ocean faces through various human activities. The challenge of dealing with these threats requires a major change in human behavior, as is expressed by researchers, artists, and the UN's Agenda 2030 for sustainable development. To meet these challenges, the community must work across many academic disciplines, using transdisciplinary approaches and developing new skills for conversation. Chapter 12 discusses the need to connect science and art, providing knowledge of societal and human values necessary to promote change. To illustrate this, Part II continues in a dialogue between a marine scientist and the ocean. The ocean sends dream images to the researcher, who interprets them based on how to connect analytical thinking and intuition, earlier described in Part I of the book. We illustrate how science and the arts can connect to generate something greater.

The dialogue begins with a dream image of a dead seabird full of plastic. The senseless killing of a seabird is the theme of *The Rime of the Ancient Mariner* by Samuel Taylor Coleridge, which illustrates how the killing of an albatross led to isolation and death. The researcher interprets the image of the dead seabird as a metaphor for the dysfunctional behavior of humans and where environmental problems can be seen as mental problems. People need help thinking more broadly about the ocean to change their behavior. The dialogue with the sea is carried forward from many different marine science perspectives.

A. Omstedt, *A Philosophical View of the Ocean and Humanity*,
https://doi.org/10.1007/978-3-031-64326-2_26

The first topic is sea ice dynamics and thermodynamics related to the threat of global warming. Should people be afraid and ashamed of using fossil fuels that have caused ice melting? The ocean responds by challenging us to cultivate our curiosity, which is a much better attitude to the ocean than fear and guilt. A major threat to the ocean would be if global warming also reduced oxygenation in the deep ocean. The ocean responds with an image of a kelp forest—among the most beautiful and biologically productive marine ecosystems and a species that has served the ocean for millions of years. Interpreting the kelp forest as a metaphor suggests a need to listen better and recognize that humans can similarly develop society as a rich habitat for human growth.

The ocean never rests. Currents, waves, eddies, and turbulence change the conditions for life on different time scales in a beautiful and complex way. Human impacts are visible in the waste, including plastic, from various sources transported across the ocean in its currents and eddies. Some believe that if we don't change our behavior, the ocean will eventually contain more plastic than fish. Instead of pristine ocean eddies, plastic accumulates by the currents into large garbage mountains that are dangerous for seabirds and marine ecosystems. The ocean's response to this risky behavior reminds us of our wasteful lifestyles and the need to improve our mental health. The ocean and the humans are vulnerable, and people need inspiration to change their behavior.

The heat balances of the ocean, land, and atmosphere determine the earth's temperature and represent an interesting and complex interplay between the sun, the earth, and our behavior. The ocean plays many important roles, such as the earth's most important heat storage. Evaporation from the ocean surface due to latent heat flux acts as a steam engine and drives large-scale wind systems. Long-wave radiation emitted from the earth's surface is partially reflected from the atmosphere by the greenhouse gases covering it and protecting it from cooling. The human impact comes from emissions of greenhouse gases, which affect the long-wave radiation that is reflected to the surface. Today, it is clear that humans are affecting the ocean through global warming and acidification of the ocean. Will humans be able to reduce global warming or not? The sea sends an image of a jellyfish swimming slowly to the surface. The jellyfish symbolizes the potential of man's inner resources: intuition, dreams, and emotions. These abilities must be used to create a better relationship between the ocean and humans for our survival.

Most of the world's freshwater is found in the ocean. Salt comes from rocks on land, is dissolved by weathering and volcanic activity in water, and is transported by rivers to the sea. Water balance and salinity are closely linked through evaporation and precipitation. Freshwater and salt are the two most important factors for life. The worried scientist asks the ocean how modern humans can have a new and better relationship with the sea. The ocean indicates that the best way is to realize that what is happening in human bodies is also happening in the sea. Humans and the ocean belong to each other through many important processes from the beginning of life.

Oxygen levels in the ocean are decreasing, causing the spread of oxygen-poor areas. The question is how to protect the ocean from dying from lack of oxygen. The ocean sends an image that looks like an octopus, perhaps one of the smartest animals on earth and with a different kind of intelligence than humans. Can people learn something new from the ocean and think more freely? The dialogue concludes that the smart solution is to establish partnerships.

Plankton provides extraordinary services by absorbing carbon dioxide and releasing oxygen, as well as serving as the base of the marine ecosystem. The connection to the ocean in this chapter is provided by the epic poem *Kalevala*, which inspires us to examine our sources of knowledge and reminds us that humans were born in the ocean. We learn that plankton is the basis of life in the ocean and that storytelling and dreams form the basis of health for us.

Human-induced declines in marine species date back more than 1000 years but accelerated markedly in the 1950s. Beautiful living marine resources are mismanaged as if humanity and the ocean no longer have a relationship. The ocean responds with dismay at human lifestyles and attitudes, urging us to listen better and realize that marine resources can return much more if managed properly.

The mineral oil formed by phytoplankton illustrates the close relationship between living and non-living resources and is today's most important fossil energy source. There is an intensive search for new sources of oil and gas. The increased interest in offshore exploration for oil and gas, also in deep-sea areas, entails many environmental risks. Will people's hunger for oil and gas destroy the ocean or be managed environmentally and sustainably? The ocean reminds us that light and darkness are deep within us all and that we can easily give in to destructive desires. A mental shift is needed, encouraging people to work together and create global cooperation and hope.

Earth's population is increasing rapidly, with many people living along coasts and in cities. Before the nineteenth century, the world's population was less than a billion; today, it is 8 billion and still growing. Will civilization collapse if we humans do not address the difficult global challenges? The ocean reminds us that we need a common vision. With Agenda 2030, the UN formulates a vision for sustainable development whereby we can develop a much better relationship with the ocean and ourselves.

How carbon dioxide could affect the air temperature was understood at the end of the nineteenth century. It was not until the late 1980s and the first assessment report of the IPCC, published in 1990, that concern about global warming spread to a larger group of people. The ocean is under pressure from many factors, and a healthy ocean is required to meet these threats. The image of a healthy ocean suggests a similarity between the sea and humans, for whom health can be physical and mental. The ocean challenges us to build health and feel joy over it, which can create something greater: harmony between us and the ocean.

An important issue for society is developing a greener and more sustainable future. The ocean responds by sending an image of barnacles on a rock. The image is interpreted as people needing to slow down, learn more from the ocean, and build sustainable and trusting relationships. Facts are not the only thing influencing attitude change; emotions are just as important. Insight into the beauty and vulnerability of the ocean and humans are factors that can inspire improved mental health and new sustainable relationships.

Part III
Science, Art, and Inspiration

Chapter 27
Opening, Part III

Abstract In the face of great human challenges, more and more people understand that examining the connection between nature and human behavior is necessary, especially what controls destructive behavior and how it can be changed. Here, we investigate the processes that may support progress and related anti-goals that lead to societal deterioration. The Greek myth of *Pandora's Box* illustrates destruction, which is then reflected in light of the UN Agenda 2030 for sustainable development. The latter can represent humans' hope that Pandora was missing when examining her box. Cooperation and listening to each other can instead generate a *Box of Opportunities* that gives us a flow of curiosity, vision, courage, hope, presence, simplicity, inspiration, and hope for the future.

Philosopher Jonna Bornemark (1973–) developed the idea that humans live in a sphere of non-knowing and must navigate with judgment (Bornemark 2020). We have seen this during the COVID pandemic when a completely unknown virus struck with a global pandemic. Uncertain knowledge and dramatic events force us to use the underestimated and overlooked human judgment.

In the face of great human challenges, more and more people understand that examining the connection between nature and human behavior is necessary, especially what controls destructive behavior and how it can be changed. Approaching an understanding of human behavior has been discussed for a long time, and natural scientists describe this in their models in a very simplified manner by prescribing human emissions under various assumed conditions, where lifestyle and society's development are important factors. In David Attenborough's (1926–) film *Life on Our Planet*, which has been shown on Netflix, he illustrates the evolution of the earth from his birth to the present day. The film is a testament to David Attenborough's studies of nature during an extreme period in earth's history with large population growth, increased greenhouse gases, dramatically reduced wilderness, and reduced biodiversity. Attenborough notes that humans are the most intelligent animals on earth. Still, if man is not to destroy himself, wisdom must increase, and only then will we have the opportunity to restore the earth's environmental balance.

A. Omstedt, *A Philosophical View of the Ocean and Humanity*,
https://doi.org/10.1007/978-3-031-64326-2_27

In the Greek myth of *Pandora's Box*, we read how the gods created a woman with the mission to punish humanity. She was given a box with her which was not to be opened. But Pandora was curious and wanted to explore the unknown. What's in the box? We all struggle between being in the safe zone and discovering and exploring new sides of life. So did Pandora, who couldn't resist looking for what was hiding in the forbidden box given to her by Zeus. She opens the box and unleashes all the world's misfortunes and diseases (Fig. 27.1). The misery can go off and move completely freely from being kept in a limited area. It is easy to imagine that Pandora was afraid and overwhelmed by difficult emotions such as shame, guilt, and anxiety. Chances are good that she would refrain from looking into the box a second time. Without examining the inside, she misses the opportunity to discover hope through the butterfly still hidden in the box, here in the figure along one edge.

Just like Pandora, we want to protect ourselves from the unpleasant. By distancing ourselves from what scares us, there is also the possibility of creating dissonance between thought and action in our lives, to the extent that it also affects our self-image. To trust that there is hope, despite our challenges, we need alternative solutions that help us look into initially uncomfortable spaces, like what humans do with the ocean and ourselves. If we bring this story to the present, the opening of Pandora's box may represent the human behavior that has created climate change, pandemics, wars, overfishing, and species decline. It awakens in us feelings such as shame, isolation, alienation, chaos, bullying, inequality, injustice, hatred, evil, and other destructive emotions.

In her famous book *Silent Spring*, Rachel Carson quoted Albert Schweitzer, who wrote: "*Man has lost his ability to foresee and prevent. She will end up destroying the earth.*" Perhaps the UN's 17 goals within Agenda 2030 can symbolize humanity's hope today, illustrated by the butterfly in Pandora's box. Table 27.1 lists the UN's global goals for sustainable development and, at the same time, the threats to them, what we want to strive toward, and what we want to aim away from. Research on the UN's Agenda 2030 allows us to deepen our understanding of human behavior and values and how they affect the physical world. Together with improved communication and transdisciplinary initiatives, this can be crucial to better understanding the interaction between the ocean and humans and creating a new relationship with the ocean.

But is the story of *Pandora's Box* a fair description of humanity today? Those of us participating in different research, literature, or dream groups notice something completely different. Perhaps we can call it the *Box of Opportunity,* which, when opened instead, gives us a flow of curiosity, vision, courage, hope, presence, simplicity, and inspiration. There is great potential for change when we learn to cooperate and listen to each other. The third part of this book is about the *Box of Opportunity.*

Fig. 27.1 Pandora, after she has opened the box and released all the world's misfortune and disease. *Illustration* Karin Hoppe Storck

Table 27.1 Goals supporting progress and related anti-goals that lead to deterioration in society

Processes of progress		Processes of deterioration	
Goal 1	No poverty	Anti-goal 1	Poverty
Goal 2	Zero starvation	Anti-goal 2	Starvation
Goal 3	Good health and well-being	Anti-goal 3	Illness and distress
Goal 4	Quality education	Anti-goal 4	Poor or no education
Goal 5	Gender equality	Anti-goal 5	Gender discrimination
Goal 6	Clean water and sanitation	Anti-goal 6	Dirty water and unsanitary conditions
Goal 7	Affordable and clean energy	Anti-goal 7	Expensive fossil and dirty energy
Goal 8	Decent work and economic growth	Anti-goal 8	Exploitive work and stagnation
Goal 9	Industry, innovation, and infrastructure	Anti-goal 9	Declining industry, innovation, and infrastructure
Goal 10	Reduced inequality	Anti-goal 10	Increased inequality
Goal 11	Sustainable cities and communities	Anti-goal 11	Fragile cities and communities
Goal 12	Responsible consumption and production	Anti-goal 12	Overconsumption and overproduction
Goal 13	Climate action	Anti-goal 13	No climate action
Goal 14	Life below water	Anti-goal 14	Failing water management
Goal 15	Life on land	Anti-goal 15	Failing land management
Goal 16	Peace, justice, and strong institutions	Anti-goal 16	War and fragile institutions
Goal 17	Partnerships to achieve the goals	Anti-goal 17	Increased antagonism and alienation

The goals follow the UN Agenda 2030 for sustainable development and its opposites (Omstedt 2023)

References

Bornemark J (2020) The horizon is always there. About the forgotten judgment. Volante, Stockholm (in Swedish)

Omstedt A (2023) How to develop an understanding of the marginal sea system by connecting natural and human sciences. Earth Syst Change Changes Marginal Seas/Oceanologia 65(1):20–29

Chapter 28
Voices of the Ocean

Abstract The ocean is full of voices or sound waves that travel all over the sea. The sound is generated from natural sources such as earthquakes, breaking ice, ocean currents, and waves. Upon this comes how all marine species communicate with each other and all noise generated by human activities. We can also look at the large amount of data collected by marine scientists, ocean poetry, and art as voices of the ocean and now as metaphors. Here, the ocean voices play into our human feelings and can generate a close connection to the sea.

Sound is transported in the ocean very efficiently, and currents, turbulence, waves, breaking ice, and earthquakes can spread different sound waves (Wunsch 2022). Sound is also the dominant way many marine species communicate with each other, and anthropogenic noise adds another layer to the ocean mixture of signals (Tripathy-Lang 2023). Isaias (2023), at the University of Trieste, Italy, examined how to examine the sound of the ocean and the music to change the human relationship with the sea in a thesis called *Seamphony: Sound of the Ocean.* Music is an expression that has existed with humans for as long as we have lived and has a strong connection to our emotions. Here, there is a great opportunity to open new ways of communication between man and the ocean.

The voices of the ocean are expressed in art and through data collected over centuries by marine scientists. Pugnetti (2020) chooses to highlight the voices of the water in her presentation of the last decade's activities for the Italian Network for Long-Term Ecological Research. The measurements become voices that one can listen to and talk to. One of the voices is the measurement series of plankton, where the researchers use their bodies and voices to make the plankton data visible and real and emerge from the water.

Our knowledge of the ocean developed through what could be discovered and what could be imagined. Knowledge passed down to succeeding generations was constantly transformed in light of experience through imagery and stories. The Iliad and the Odyssey are poems recorded by Homer around 700 BC and are examples of stories that have gained great importance for western culture and the view of the ocean. Myths illustrate that the ocean is where everything is born and where

A. Omstedt, *A Philosophical View of the Ocean and Humanity*,
https://doi.org/10.1007/978-3-031-64326-2_28

everything will end, and they are filled with evocative images. The gods have been active in the human imagination even before antiquity, giving the ocean different voices. In modern times, the gods and mythology are primarily evident in the names of ships and marine instruments. The Swedish icebreakers are named after Nordic gods such as Ymer, Thor, Oden, and Frej. Of the instruments, the Argos buoys are named after a giant with many eyes who, according to Greek mythology, could see everything, while the underwater craft used to measure the ocean under the Antarctic ice is named Ran after the mother of the sea in Norse mythology.

A beautiful example of an early description of the ocean is the Carta Marina, the first map to present details and names of the Nordic countries (Fig. 28.1). The map was drawn by Olaus Magnus (1490–1557), a Swedish priest in exile in Rome, and published in 1539. It contains a wealth of information about, for example, coastlines, hunting, fishing, sailing ships, ice skating, polar bears, whales, sea lions, walruses, crabs, lobsters, giant sea snakes, and other monsters. The map gives a vivid picture of the ocean where many voices call for attention. By comparing the map with modern observations, it has been possible to show that some of the eddies depicted on the map show similarities to those observed in the sea off the Iceland–Faroe Islands.

The ocean has challenged humans to explore the unknown, leading to questions such as: What lies beyond the horizon? How deep is the ocean? Why is sea water salty? What determines and drives ocean currents? Are marine resources unlimited? Scientific curiosity has led to many research expeditions. Early expeditions linked to trade and fishing mapped new lands, dangerous coasts, islands, and straits. The Challenger Expedition (1872–1876) initiated modern oceanography by systematically collecting data and inspiring new voyages. The Vega and Albatross expeditions were the two largest Swedish oceanographic expeditions. The Vega expedition was the first expedition to cross the Northeast Passage, 1878–1880, while the Albatross expedition circumnavigated the globe in 15 months in 1947 and 1948. Since World War II, several major international expeditions and science programs have been organized to collect observations and develop knowledge about the ocean. Significant progress has been made by standardizing measurements, building open databases, and organizing international conferences. The progress develops through an intensive conversation where the voices of the ocean are connected to the researchers' many presentations and discussions.

Observations at the ocean require great commitment and knowledge of navigation and mapping. This knowledge was previously based on indirect methods such as observing driftwood, stars, and seabirds, studying written documents from various archives, and listening to the sailors' stories. Later, but before the electronic era, the sextant was developed for safer navigation and several mechanical current gauges. Other important instruments were thermometers that could measure temperatures at different depths and the Nansen bottle that could collect water samples from different depths. At the same time, new instruments were built to measure water levels, sediment samplers, and various nets to catch plankton and fish. Many scientists were academically trained and became inventors after working closely with technically skilled people. Expeditions on ships also trained marine scientists in interdisciplinary collaboration.

Fig. 28.1 Carta Marina was the first map to present details and names of the Nordic countries. The map was drawn by Olaus Magnus, a Swedish priest in exile in Rome, and published in 1539

Marine science has developed through the constant interaction between theory, observations, and experiments in laboratories and at the ocean, through people's curiosity and deep commitment. The three main elements that distinguish ocean mechanics from traditional fluid mechanics are the effects of rotation, stratification, and turbulence. The theoretical considerations often use mathematics based on classical mechanics to derive the geophysical equations of motion. These complex, nonlinear equations require insights from observations and good intuition. The observations led to theoretical successes. For example, Leonardo da Vinci (1452–1519) illustrated as early as around 1507–1509 that turbulence involves eddies on different scales by drawing eddies in a bathtub.

Carrying out measurements in the ocean is difficult for various reasons (Wuench 2015):

- Seawater is corrosive and attacks metals chemically,
- seawater is opaque to electromagnetic radiation,
- seawater causes biological fouling of instruments,
- and the instruments are exposed to high pressures at great depths.

In addition, conditions in the ocean vary greatly over space and time, creating intense impacts through waves and ice. Technological developments, often driven by the military need for ships, navigation systems, satellites, and new instruments, have changed oceanographic measurement methods significantly over the past century. Today, robotic devices such as Argos buoys drift with currents and move up and down the ocean, automatically measuring and transmitting this data to satellites. Other new measuring instruments include self-propelled or remotely operated submersibles and bottom pressure gauges. But some areas are very difficult to explore, for example, the Polar Seas, where the ice causes great problems. At the same time, these unknown areas are central to our understanding of the climate. Scientists have creatively begun collaborating with elephant seals in the hard-to-reach ocean areas around Antarctica. Various sensors are attached to the elephant seals, which, for a long time, can dive at the ice edge and under the ice, giving a completely new picture of what was previously hidden (Biddle and Swart 2020).

Despite great successes, the ocean is still, in many respects, unknown and harbors many hidden resources. Important problems remain to be solved, mainly how humans affect the ocean. The voices of the ocean in the form of observations, sounds, and artistic creations can powerfully contribute to a system change that works against antagonism and alienation.

References

Biddle LC, Swart S (2020) The observed seasonal cycle of submesoscale processes in the Antarctic marginal ice zone. J Geophys Res: Oceans 125:e2019JC015587

Isaias EA (2023) Seamphony: sound of ocean; by Marios Joannou Elia. Thesis, Advanced master in sustainable blue economy, 5th edn. National Institute of Oceanography and Applied Geophysics—Trieste, and University of Trieste, 66 pp

Pugnetti A (2020) Voices from the water: experience, knowledge, and emotions in long-term ecological research (LTER Italy). Adv Oceanogr Limnol 11:59–70

Tripathy-Lang A (2023) Oceanic cacophony. Eos 104. https://doi.org/10.1029/2023EO230500. Published on 21 December 2023

Wuench C (2015) Modern observational physical oceanography: understanding the global ocean. Princeton University Press, Princeton, NJ

Wunsch C (2022) Can oceanic flows be heard? Absyssal melodies. J Acoust Soc Am 152(4):2160–2168. https://doi.org/10.1121/10.0014603

Chapter 29
Scouting Beyond the Horizon

Abstract This chapter will illustrate how humans affect the Baltic Sea, one of the world's most studied sea areas with many climate and environmental problems. The many human impacts are divided into naturally occurring factors that humans influence (e.g., climate change) and factors linked to human activities (e.g., fishing). A large mixture of human impacts threatens the Baltic Sea and many coastal seas. Many of them are connected, and today, there is a concern that we need to understand more about how these factors work together and amplify or counteract the impact on the sea.

Developing knowledge is not enough to solve problems; one should also be able to anticipate new problems requiring new knowledge. The history of the ocean under human influence illustrates that the ocean has entered a completely new phase and has changed greatly in recent millennia. We face a choice between two very different oceans: one healthy and productive, managed sustainably, and another overexploited and continuing to decline (Rogers 2019). Examples of countermeasures that can lead to a healthier ocean are reversing the management paradigm. Today, marine reserves cover only a small part of the ocean. The rest is intensively exploited. Instead, a healthy ocean needs extensive marine protected areas, estimated to be 20–40% of the ocean area. Another example is the UN's Ocean Decade for Sustainable Development, 2021–2030. This decade is dedicated to developing scientific knowledge, shared and enhanced information systems, science-based fisheries, and behavior change.

Here, we will illustrate how humans affect the Baltic Sea, one of the world's most studied sea areas with many climate and environmental problems (Fig. 29.1). Failed management of cod and Baltic herring show the distance between knowledge and management. The following description is based on an attempt to depict the entire matrix of relationships between human influence and its effects on the Baltic Sea (Reckermann et al. 2022). This evaluation was compiled in 2022 by a research group within Baltic Earth. Roughly, one can distinguish between naturally occurring factors that humans actively influence and factors solely linked to human activity. Examples of the first category are

A. Omstedt, *A Philosophical View of the Ocean and Humanity*,
https://doi.org/10.1007/978-3-031-64326-2_29

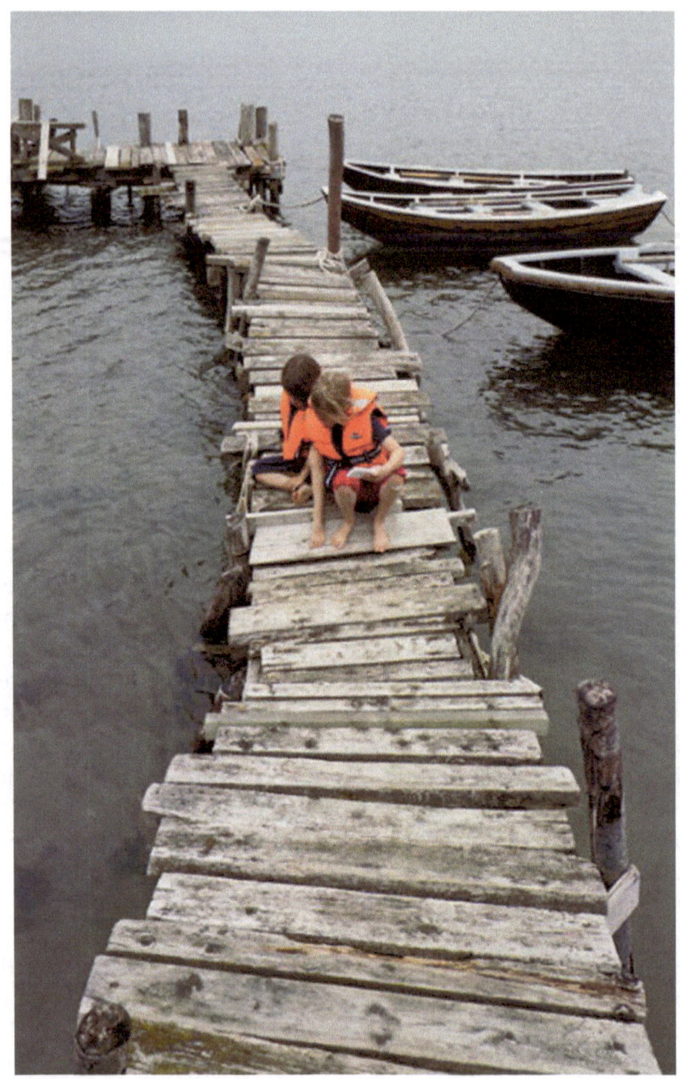

Fig. 29.1 How long will the bridge last? How long will our society last? *Photo* Hillevi Nagel

- Climate change,
- warming,
- acidification,
- coastal processes,
- oxygen-poor sea areas,
- groundwater discharge,
- marine ecosystems, alien species,
- and land use.

The latter category, where the problems have arisen solely through human activity, includes

- Urban and coastal construction,
- marine wind farms,
- shipping,
- fishing,
- chemical contamination and drug residues,
- dumped ammunition,
- marine debris,
- production and consumption of plastics, including microplastics,
- agriculture,
- water regulation,
- tourism,
- and coastal management.

These factors are, in turn, driven and influenced by

- Food production,
- energy production,
- transportation,
- industry,
- and other economic activities.

Many of them are connected, and today, there is a concern that we need to understand more about how these factors work together and amplify or counteract each other. It is about understanding how to manage a sustainable relationship between humans and the sea, where many interests are included. Of course, factors such as population growth, urbanization, technological development, lifestyle, and values also play a significant role. All these different factors and interactions create a need for strong political leadership and successful administration. However, the starting points cannot be solely to optimize various human claims to utilize the marine resources in a technically optimal way. But also a quest for harmony and beauty between humans and the ocean as a guarantee for a healthy future. Here, something more is required than traditional science's offer of knowledge facts. Instead, we need a broader perspective.

What constitutes knowledge has been constantly debated in philosophy. Classical philosophy of knowledge takes its starting point from Aristotle (384–322 BC), who divided knowledge into three parts (Gustavsson 2000):

- Episteme,
- teche,
- and phronesis.

Plato (428–348 BC) established the idea of certain knowledge, episteme, that cannot be otherwise, as in logic and mathematics, instead of merely having an opinion or thought about something. Aristotle broadened the understanding of knowledge by formulating the concept of techne for the knowledge we need in practical activities, such as craft or art, which concern anything humans produce, manufacture, or create. In further development, certain knowledge, episteme, became the designation for scientific knowledge. But to live a good life, we also need knowledge of interpersonal relations. We need such social or dialogical knowledge for social and political action, which Aristotle calls phronesis: practical wisdom or good judgment. Phronesis means performing the right action at the right time and doing it as well as possible. It is the judgment we need when we do not know everything but still have to act. Through experience and intuition, this judgment is built up and leads skilled people to navigate uncharted waters. The next chapter examines which sources of knowledge are at our disposal.

References

Gustavsson B (2000) Philosophy of knowledge. Three forms of knowledge in historical light. Wahlström & Widstrand. ISBN 978-91-46-17648-0 (in Swedish)

Reckermann M, Omstedt A, Soomere T, Aigars J, Akhtar N et al (2022) (2022): Human impacts and their interactions in the Baltic Sea region. Earth Syst Dynam 13:1–80

Rogers A (2019) The deep: the hidden wonders of our oceans and how we can protect them. Wildfire, Headline Publishing Group, London, UK

Chapter 30
Thinking and Sources of Knowledge

Abstract How we see and perceive reality depends greatly on our personality and background. Some see more and more details (atomistic approach); others emphasize that the whole is more than the sum of the details (holistic approach). A similar pair of concepts is particular and universal. The particular can be expressed in parts like a beach, nation, or city. The universal is the whole, which generally applies to the ocean or the UN Charter. A photograph in the chapter illustrates how one can think in different ways. It analyzes what one can see if one puts on scientific, artistic, or mythological glasses. Bringing the different views together allows many perspectives and provides great power and inspiration.

How we see and perceive reality depends greatly on our personality and background (Omstedt and Gustavsson 2022). Some see more and more details (atomistic approach); others emphasize that the whole is more than the sum of the details (holistic approach). A similar pair of concepts is particular and universal. The particular can be expressed in parts like a beach, nation, or city. The universal is the whole, which generally applies to the ocean or the UN Charter. In the natural sciences, we talk about process and system understanding. The three pairs of concepts, atomistic-holistic, particular-universal, process, and system understanding, can be illustrated in diagrams (Fig. 30.1). Marine science increasingly goes into detail when, for example, new instruments are developed. Developing these instruments and analyzing the results they generate often require extensive specialist expertise and can change how we see the world.

The connection between process and system understanding was illustrated during the Nobel Prize in Physics 2021. The prize-winning work focused on the complexity of physical systems, ranging from the microscopic structure of glass to human influence on the climate. The award justification was formulated as a "*groundbreaking contribution to our understanding of complex physical systems.*" Half the prize went to Syukuro Manabe and Klaus Hasselmann, and the other half to Giorgio Parisi. Manabe showed how increased carbon dioxide content in the atmosphere gives rise to higher temperatures on the earth's surface due to the interaction between the radiation balance and the vertical transport of air masses. Hasselmann created a model that

A. Omstedt, *A Philosophical View of the Ocean and Humanity*,
https://doi.org/10.1007/978-3-031-64326-2_30

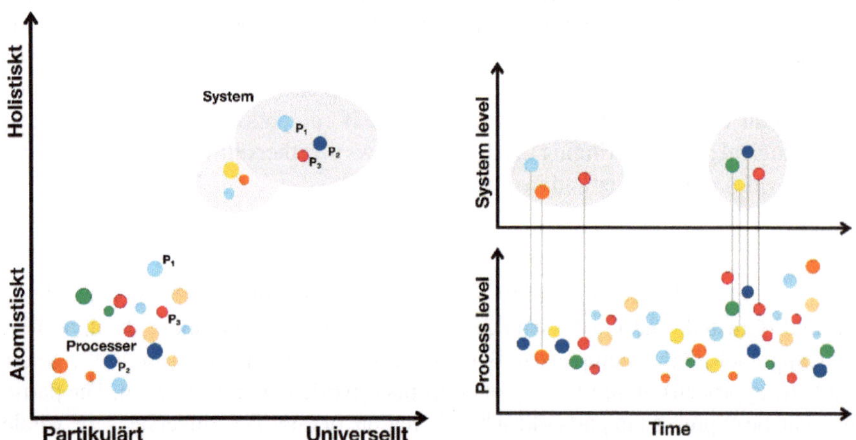

Fig. 30.1 Sketches of seeing the world from different perspectives with circles indicating processes, some of which are parts of a system (gray). Our beliefs also change over time (right diagram). *Illustration* Jan Heuschele redrawn from the article by Omstedt and Gustavsson (2022)

linked weather and climate, showing how climate models can provide information even if the weather is variable and chaotic.

Atomistic and holistic, particular and universal, or process and systems understanding should not be treated as contradictory ideas; when these complementary perspectives meet, they can transform our understanding of reality and each other. But with an atomistic view, there is a risk that knowledge and treatment of nature and people can be split into many separate and conflicting compartments. The holistic view instead opens up the whole but with the risk of simplifications. Our understanding of each other and the world also changes over time, where new knowledge opens up for increased understanding. For example, our understanding has evolved from a geocentric to a heliocentric worldview through Newton's understanding of nature, and later to the insights of relativity and quantum mechanics, or from Linnaeus' classification of species to Darwin's studies of evolution, and to today's DNA studies; or from pre-industrial societies, through the industrial revolution and now to post-industrial societies.

While we approach a problem, humans have different internal interpretation patterns. This can be illustrated if we look at the photograph in Fig. 30.2. From the knowledge source of science, we see a wall, a meadow, a sea, and a sky. If we go deeper, we can determine the time of the wall's construction, the season from the light, and the plants in the meadow. Scientific thinking can then be led into ever deeper details about processes that determine, for example, how the different parts affect each other through flows of heat, water, and gases such as oxygen, carbon dioxide, and methane. The scientific source of thought is problem-solving and detail-oriented and can solve many questions, but it also entails a risk of one-sidedness and deadlocks.

If we instead study Fig. 30.2 from our sensuous and artistic source of thought, we will be emotionally affected. Emotions such as joy and wonder at the picture's beauty come to us. The creative source of thought opens our feelings and can give us inspiration that cannot be expressed in words. It also has a sympathetic feature in that it is ambiguous and therefore dependent on the viewer himself, and thus, what is right and wrong does not prevail here, but the interpretation is free.

Let's go further to the mythological source of thought with the fairy tales, religions, and the fantasy genre of literature. We discover a person stuck in the wall and locked, but with inspiration, that is, the opening, the sea, and the light behind him. Is liberation possible? In the mythological source of thought, there are often unexpected solutions, as in, for example, the H. C. Andersen (1805–1875) tale of *Dummerjöns*, who, against all odds, got the princess. Another tale from H. C. Andersen is about *The Emperor's New Clothes*, where the child sees through the bloated king, and the tale teaches us that the truth can be found in the most unexpected, the child. In the same way, the arts and dreams are bearers of the deepest truths about us humans, which can lead us in new directions and free us from blockages. The knowledge we find in the artistic and mythological source of thought is separate from the scientific source and has a different logic and grammar. For example, the criterion of falsification does not apply to art and dreams because only the observer can guess the scope of the observation.

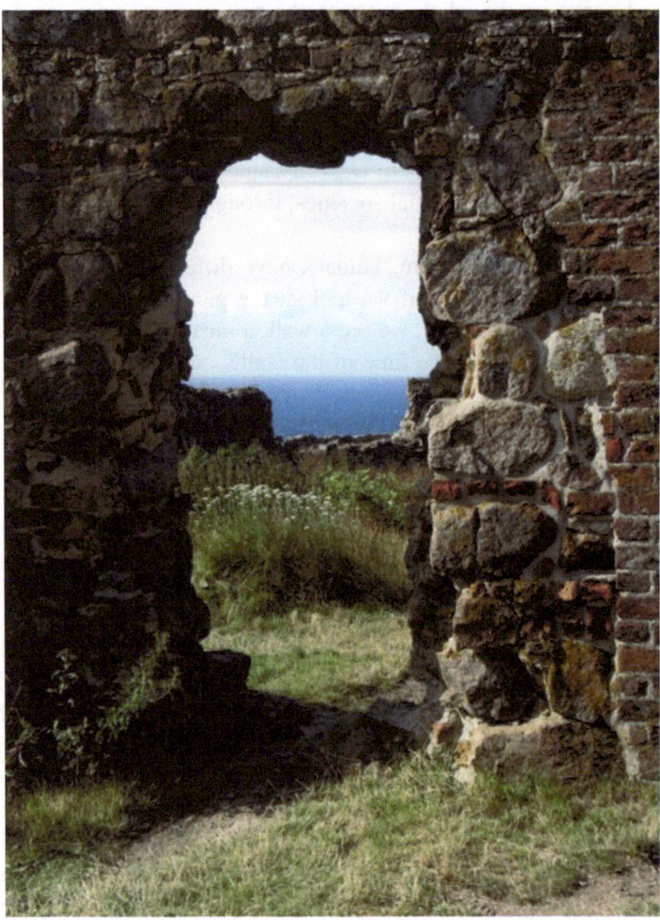

Fig. 30.2 Baltic Sea is seen from the island of Bornholm. *Photo* Anders Omstedt

Kahneman (2016) investigated fast and slow thinking, or intuitive and analytical thinking, when humans evaluate and make decisions. He states that research today shows that the intuitive thought system influences our choices and judgments more than we think. In her book *Eurydice's Song*, the pedagogue and literary scholar Stina Hammar (1923–2020) talks about the two sources of thought, the abstract and the sensual (Hammer 1997). The sensual source of thought contains the artistic and mythological. To illustrate the grammar of myths, dreams, and fairy tales, Stina Hammar analyzed texts by Lars Ahlin, Fjodor Dostoevsky and Maksim Gorky. At the bottom of her interpretation is the tale of Orpheus and Eurydice. Orpheus got his lyre from Apollo, the god of light, so the source of abstracting thought starts from the clear, logical order of the head and the words. Just as Orpheus could not free his beloved Eurydice from Hades, the abstracting source of thought does not reach the full depth of our life. The sensual source of thought belongs to Eurydice, and according to Lars Ahlin, contrary to the myth story, her wisdom (the art, the dream, the game) can free Orpheus from Hades. Life is processed and changed by uniting both sources of thought.

Each source of thought can lock us into biases, black-and-white thinking, and destructiveness, so learning how to connect these sources of thought is important. The author Salman Rushdie (1947–) told the TV program Skavlan that the main idea in his novel *Two years, eight months and twenty-eight nights* (One Thousand and One Nights) is that the rational side—science—and the irrational—dreams and imagination—together gives enormous power. But if they break apart, monstrous things happen.

Many need to develop their scientific thinking but risk getting stuck if they do not also allow themselves to be inspired by the beauty of nature and humanity. At the same time, the mythological source of thought teaches that much in life is a mystery, and no matter how good the scientific methods are, much cannot be understood. However, if the various sources of thought are brought together, hope is given for the future, a hope that is beyond all reason.

How modern brain research looks at our brain is described by Jill Bolte Taylor in a TED talk (Bolte Taylor, 2008). There, she emphasizes that the left and right hemispheres of the brain can be regarded as different personalities. The left focused on details and time, and the right focused on the whole and presence. You can say that the left hemisphere is atomistically oriented, and the right hemisphere is more holistically oriented. The important thing is that all people can access these different abilities, and both can be trained.

When the author Julia Ravanis (1993–) discusses the theoretical physics description of the universe in her book *The Beauty in Chaos*, she takes help from human experience (Ravanis 2021). Explanations of black holes, entangled electrons, and string theory are interspersed with reflections on longing, frustration, and falling in love. Despite all the questions the universe opens up, we only have ourselves to relate to.

David Bohm (1917–1992) formulated the hypothesis that behind the visible reality is an underlying order that is not directly knowable to us but nevertheless forms the basis of our existence (Bohm 1980). He articulated our strong need to see the whole because fragmentation is so prevalent in today's society, and it is an illusion that the fragmented parts exist independently. Instead, wholeness is the real thing, and humans have created the fragments through a wrong approach. For example, seeing a person as atoms, hormones, chemistry, or body parts is a limited view because each person is a whole. Correspondingly, to manage the ocean sustainably, we must be able to see and work for the whole human and ocean system. An emphasis on harmony and beauty between humans and the ocean strengthens a more holistic approach.

References

Bohm D (1980) Wholeness and the implicate order. Routledge Classics, New York, USA

Bolte Taylor J (2008) My stroke of insight is available as a TED talk on. https://www.ted.com/talks/jill_bolte_taylor_s_powerful_stroke_of_insight

Hammer S (1997) Eurydice's wisdom. A book about the sensual source of thought. Akademitryck AB, Edsbruk. Sweden (in Swedish)

Kahneman D (2016) Think fast and slow. (trans: Svensson P). Månpocket (in Swedish)

Omstedt A, Gustavsson B (2022) The complex interactions between humans and the marine environment require new efforts to build beauty and harmony. Front Mar Sci 9:913276. https://doi.org/10.3389/fmars.2022.913276

Ravanis J (2021) The beauty in Chaos. Natur & Kultur, Stockholm, Sweden. ISBN 978-91-27-17004-9 (in Swedish)

Chapter 31
Why Should We Care?

Abstract The earth has become a beautiful "*blue planet,*" supporting life in many diverse forms. The development into a blue planet surrounded by life-giving conditions has been going on for billions of years. Only in the most recent time of the earth's history have humans become a part that also affects the planet. The earth system and society always evolve dynamically, and there is no turning back. Some hope to restore nature and the sea to an earlier historical period, but this will not be possible. Instead, the future of the ocean is closely linked to human development. Therefore, the idea that the ocean takes care of itself is fundamentally wrong. The effort to restore to the way it was before is impossible. Unlike our ancestors, however, today, we have access to much knowledge that allows us to change how we misuse the ocean and instead create the future we want. This is expressed in the UN Ocean Decade *to safeguard the ocean we need for the future we want.* It is, therefore, easy to understand why we should care.

Earth is also called the "*blue planet*" because most of its surface is water. But also because the sunlight is composed of light from blue to red, where the blue light spreads more and colors the atmosphere and the ocean blue. The sea is continuous and connects the polar and equatorial regions and all the coasts of the earth. Therefore, today, we speak of the ocean as a unit and refer to it in the singular. The main components of the ocean system are

- The geosphere system—rocks, sediments, and minerals,
- the physical system—driven by the global heat and water cycles,
- the chemical system—all known chemicals and the carbon cycle,
- the biosphere system—all living marine organisms,
- and the human system—all human activity that affects the ocean.

The ocean responds to forcing with its dynamics, including currents, waves, eddies, turbulence, ice, temperatures, salinities, and ecosystems. Traditionally, marine sciences have often neglected the human component, but in the last decades, the realization of the strong human influence on the ocean has increased. Marine science has become a field with many inter- and transdisciplinary challenges. Success

A. Omstedt, *A Philosophical View of the Ocean and Humanity*,
https://doi.org/10.1007/978-3-031-64326-2_31

in meeting these challenges requires a strong collaboration between the natural sciences and the humanities and commitments from society.

For us humans, the perception of time and space is often very limited, even though we grapple with questions from the size of elementary particles to galaxies. This can be illustrated with time written as powers of ten, i.e., 10^p where p is the power and 10^0 corresponds to a year. We can often relate to changes within about 1 s to 100 years. This corresponds to the p changing from − 7 to 2 or 10^{-7} to 10^2. Chemical reactions can be fast, occurring in milliseconds, or about 10^{-10} years, while the earth is 4.5 billion or about 10^9 years old. This illustrates how processes on earth operate over a much larger time interval than humans can perceive in everyday life. Carrying out a similar thought experiment with the spatial dimension, most of us can relate to changes ranging from perhaps millimeters to kilometers or from p values ranging from − 3 to 3, the unit now being a meter. Again, if you look at the earth system, with changes from the molecular level to the earth's radius (about 6371 km), the power varies from − 10 to 7 or 10^{-10} to 10^7 m. The numbers become even larger if we consider spatial scales, including the moon, sun, and galaxies. The implication is that humans do not perceive many important processes on and for the earth in our everyday lives. Different ways of looking at reality or different "glasses" are often needed but can lead to important aspects being overlooked. One can also say that reality is much bigger and smaller than humans, but it is through a network of processes in humans' inner ocean that the understanding of reality is understood (Fig. 31.1).

At the same time, we live in an age of accelerating information flows that demand our attention. Every day, our brains must take in a lot of information and deal with the accompanying frustration. Will our lives become more fragmented, or can we find a way to become more integrated? Ullman (1996) believed that there are two ways of knowing the world and our relationship to it:

- The scientific approach,
- and the way discerned in our dreams.

He also regarded the dream as an adaptation concerned primarily with the humans' survival and only secondarily with the individual, as an inner ocean whose message comes to receptive individuals. In society, a kind of sale of facts and values constantly occurs in the triangle between politics, media, and science. For science to remain credible, researchers must make their description tenable.

Hans von Storch (1949–) listed several prerequisites for the sustainable practice of climate science (von Storch 2012). He states that researchers must recognize that scientific knowledge is uncertain and often needs future revision. One must also understand that the public narrative of climate change is driven by two different narratives, one of science and its claim to knowledge and another of the media and its claim to attention. Scientists must explain that reality is very complex. Also, accepting an intricate understanding of the dynamics of climate change means that simple policy conclusions about necessary or meaningful action cannot be drawn. Scientific knowledge has clear limitations, and in every democratic society, there needs to be a clear division of responsibility related to science and politics. But this

Fig. 31.1 Wandering albatross flying in Amundsen Sea, Antarctica. *Photo* Christian Stranne

does not always help as the political arena has many other and often short-term claims. This is illustrated in, e.g., managing fish stocks, where researchers via ICES assess sustainable fishing based on information that is difficult to estimate. This information is the basis for political negotiations where the limits for permitted fishing are often raised, a sad example of the tragedy of the commons, where individual interests lead to problems for the commons.

The question of knowledge is not only a question for science but also for all aspects of being human. In our time, where the concept of truth has been relativized, it can help to examine how we have previously related to the concept of truth. Gustavsson (2021) explored this from different philosophical perspectives. The author argues that understanding and explanation must be combined between the humanities and natural sciences, including literature, art, and ethics. In the book, he refers, among other things, to a statement from the author Karl-Ove Knausgård (1968–), who says that literature is not truth but an area where truth is shaped. Truth is not one-size-fits-all but often requires many perspectives to meet. In the same way as Knausgård, we can say that art and dreams are not the truth but can carry the observer's deepest truth.

So why should we care? Earth's development into a blue planet surrounded by life-giving conditions has been going on for billions of years. Only in the most recent time of the earth history have humans become a part that also affects the ocean. The earth system and society always evolve dynamically, and there is no turning back. Some hope to restore nature and the sea to an earlier historical period, but this will not be possible. Instead, the future of the ocean is closely linked to human development. Therefore, the idea that the ocean takes care of itself is fundamentally wrong. The effort to restore to the way it was before is impossible. Unlike our ancestors, however, today, we have access to much knowledge that allows us to change how we misuse the ocean and instead create the future we want. This is expressed in the UN Ocean Decade *to safeguard the ocean we need for the future we want*. It is, therefore, easy to understand why we should care, which is also the theme of this book.

References

Gustavsson B (2021) The pursuit of truth in the era of postal truth. Ekström&Garay, Lund (in Swedish)

Ullman M (1996) Appreciating dreams: a group approach. Sage Publications Inc. International Educational and Professional Publisher. Thousand Oaks, California, USA

von Storch H (2012) Sustainable climate science. I: Reckermann M, Brander K, MacKenzieoch B, Omstedt A (red.) Climate impact on the Baltic Sea: from science to policy. Springer-Verlag, Berlin, pp 201–209

Chapter 32
Literature and Storytelling as Inspiration

Abstract Literature and stories create a prerequisite for understanding other people's living conditions and feelings and are not only about the past but also about the present and future. Literature searches for the truth about the future as the natural scientific climate models do in the various conceivable future calculations and scenarios, also called storylines. The UN's Ocean Decade aims to break the negative trend that is taking place in and around the ocean today and to create the conditions for sustainable development. A paradigm shift is required, from exploiting the ocean in an unsustainable way to serving the sea, that is, uniting humans and the ocean in a new relationship, similar to when King Odysseus and Queen Penelope join after many years and form the perfect union that guarantees vitality and progress. *The Odyssey* by Homer, one of the greatest adventures in world literature, can be read as a metaphor for a human's violent lifestyle, and Penelope as a metaphor for the ocean's resources. Only when these poles can meet and unite are conditions for sustainable development possible.

Many of us try to learn from our conscious and unconscious dimensions, and we all know that we can never fully understand what lies in the unknown or the future. New knowledge can be elicited and understood differently, and many methods are available to promote this exploration. Science, literature, art, music, and dreams all try to illuminate important aspects of being human. The playground of creativity is where our outer and inner dimensions meet. Of course, hard work is required, and many who explore their subjects in great depth sooner or later need to work on their personality. Encountering oneself through others, understanding one's vulnerability, and building good memories are gateways to humanity. Dreams, art, and literature can teach us much about these processes. The author Fyodor Dostoevsky (1821–1888) describes at the end of his book *The Brothers Karamazov* how one of the brothers gave a speech to boys who had just buried one of their young friends. The speech expresses the importance of having good memories, especially from an early age, and that even if a person has only one good memory, they are saved.

Often, we only remember times when things went wrong and when we or others failed. In his last novel, Dostoevsky, like later authors of more recent literature, for

A. Omstedt, *A Philosophical View of the Ocean and Humanity*,
https://doi.org/10.1007/978-3-031-64326-2_32

example, Svetlana Aleksievich (1948–), emphasizes the importance of good memories as the basis for human growth. In Aleksievich's book *Prayer for Chornobyl. Chronicle of the Future* illustrates the great importance of literature in bringing to life the relationship between humans and technological development. The book describes the people and animals affected by the very serious reactor accident at the Chornobyl nuclear power plant, north of Kiev, Ukraine, on April 26, 1986.

Literature and stories create a prerequisite for understanding other people's living conditions and feelings. *The Odyssey* by Homer, one of the greatest adventures in world literature, is an example. It began as an oral story and was later written down, perhaps by Homer, about 700 years before our era. This story has been interpreted and rewritten in many ways. It is a symbol and a starting point for a large part of our culture created through literature, art, and philosophy. In philosophy, it is primarily the very structure of departure and return that has developed. The first step is belonging, the beginning, in what is already familiar, in what we recognize ourselves in, and from which we start when we encounter something new, different, and exciting. The second step is the journey itself, to leave home, the familiar and accustomed, and open oneself to new experiences, to expose oneself to risk, to forget oneself as in play, and to open oneself to other interpretations that lead to a new understanding. The third step consists of the return itself, returning to the home we left and then coming to a new home with new experiences. The homecoming, the third step, is essential in the formation process. It is not putting oneself at risk but the return home that constitutes the essence of education. The story has been interpreted in many ways, giving rise to other stories.

An example is Eyvind Johnsson's *The Swell of the Beaches*, written during the Second World War and published in 1946. Here, Eyvind Johnsson (1900–1976) lets Odysseus come forward with feelings of resignation and fatigue from his constant struggle. The longing for home, doubts about his tasks, and the decision to take his place shines through the story. Another example is the author Margret Atwood (1939–), who raises the female perspective in her book *The Penelopiad*. Odysseus' wife, Penelope, who sits at home in Ithaca, waits for her husband's return and manages their home, surrounded by suitors who eat her out of the house.

Literature is not only about the past but also about the present and future. Literature carries the same search for the truth about the future as the natural scientific climate models do in the various conceivable future calculations and scenarios, which are also called stories. In the literature, predictions are based not on mathematical expressions but on human experience and creativity.

The Odyssey illustrates how people are trapped in a life ruled by the gods, where violence begets violence beyond human control. Today, we act as if we are similarly deprived of agency, and this sense of powerlessness breeds violence, which manifests itself in destructive behavior toward the ocean, among other things (Omstedt and Gustavsson 2022). We are caught up in short-term consumption thinking and do not take responsibility for our behavior toward nature and each other. The UN's Ocean Decade aims to break the negative trend that is taking place in and around the ocean today and to create the conditions for sustainable development (Pendleton et al. 2020). A paradigm shift is required, from exploiting the ocean in an unsustainable way to serving the ocean, that is, uniting humans and the ocean in a new relationship, similar to when King Odysseus and Queen Penelope join after many years and form the perfect union that guarantees vitality and progress. Here, Odysseus can be seen as a metaphor for a human's violent lifestyle, and Penelope as a metaphor for the ocean's resources. Only when these poles can meet and unite are conditions for sustainable development.

From Odysseus' ten-year journey, we can be inspired to reflect on what is needed for a successful Ocean Decade (Omstedt and Gustavsson 2022). The story shows the need for factors such as

- Practical and wise leadership,
- clear visions,
- community involvement,
- information exchange,
- recognition of the vulnerability of both the ocean and people,
- and shared stories.

Perhaps we must also reflect on what the ocean communicated in Chap. 20, which says that storytelling and dreams are as important to human health as plankton is to the ocean.

References

Omstedt A, Gustavsson B (2022) The ocean: excursion and return. Filosofia (67). https://doi.org/10.13135/2704-8195/7241

Pendleton L, Evanse K, Visbeck M (2020) We need a global movement to transform ocean science for a better world. PNAS 117(18):9652–9655

Chapter 33
Summary, Part III

Abstract The book's third part extends the discussion about the need to connect science and art as inspiration for behavior change. Here, *Pandora's Box* is opened as an illustration of destruction and relates to goals supporting global progress and related anti-goals that lead to societal deterioration. The alternative is the *Box of Opportunity,* which, when opened instead, gives us a flow of curiosity, vision, courage, hope, presence, simplicity, and inspiration, which has a great potential for change when humans learn to cooperate and listen to each other. The third part takes the reader through inspirations from ocean voices, humans' impacts on coastal seas, philosophy, why we should care, literature, and storytelling. The aim is to illustrate that a fundamentally new way of thinking is possible, where humans and the ocean are united in a new relationship that guarantees vitality and progress.

Today, we face many challenges, and navigating safely into the future requires knowledge and judgment. David Attenborough states that if humans are not to destroy themselves, wisdom must increase, and only then do we have the opportunity to create harmony between the earth and humans. The limitations of the ocean have become clear in many ways, for example, through the fact that many areas are overfished, but also through the presence of plastic in the ocean, which poses a great danger to marine life. The climate changes caused by human's extensive mineral oil and gas consumption are becoming increasingly apparent. We are faced with a choice. One path leads to increasing ocean degradation, the other to a completely new relationship between humans and the ocean. Agenda 2030 and the Ocean Decade (2021–2030) are international UN initiatives highlighting the need to change human behavior. And this change depends on how we think and collaborate in new ways.

How we see and perceive reality depends a lot on our personality, culture, and what we are trained in. Sometimes, an atomistic approach dominates, and details become increasingly important. Others see more of the whole or emphasize holistic thinking where man and nature are one. In society, the individual or the particular is stressed. Its opposite, the universal, however, is becoming increasingly important as many human activities touch the entire earth. Within science, there is interest in understanding processes, but system understanding is also required to understand

A. Omstedt, *A Philosophical View of the Ocean and Humanity*,
https://doi.org/10.1007/978-3-031-64326-2_33

the whole. That people see reality differently is not a problem, provided that all perspectives can be heard in open conversations. Therefore, thinking broadly and pluralistically is a prerequisite for arriving at a new approach.

At the same time, humans have access to both rational and subjective thinking. One can also speak of analytical and intuitive thinking, scientific and artistic, or facts and feelings. Intuitive or creative thinking includes dreams and imagination. When science and art are brought together, new possibilities for new ideas open up. Today, more and more people realize that there is no contradiction between facts and feelings but that they are instead closely connected and together give humans great problem-solving potential. It should, therefore, be a matter that all education contains scientific and artistic elements.

As humans, our perception of reality is often very limited in time and space. Modern physics opens up breathtaking perspectives in time and space. It is, therefore, no wonder that it needs art to describe the world. An interesting example is when Julia Ravanis uses human experience to describe theoretical physics. In human thought, the search for truth is not a matter for science alone but also for all aspects of being human. Literature and dreams can be seen as areas where the personal truth is shaped, often into stories of great value. In the encounter with science, these stories can better develop our thinking and lead to changed behavior.

So why should we care? The development of the ocean has been going on for billions of years and has created a blue planet rich in many aspects that have made life possible. The ocean's development has taken place without our interference. Only recently has human influence become part of the ocean's development. The future of humans and the ocean are closely linked. That the ocean can take care of itself is a wrong thought. It is also impossible to restore the ocean to its former state because the conditions have changed. Today, people have access to knowledge completely different from earlier generations. It allows us to improve our behavior toward nature and create our desired future. It is, therefore, easy to understand why we should care and make major changes in our relationship with the ocean. But also the need to strengthen the human relationship with all life on earth. Examples of processes leading to deterioration are listed in Table 1 in Chap. 27. It is easy to see that we are facing major challenges and why the universal values formulated in Agenda 2030 are central to human progress. Our access to scientific and artistic sources of knowledge inspires hope and insight into the possibility of a fundamentally new way of thinking, where humans and the ocean can be united in a way that guarantees vitality and progress.

Book Summary

Humans are connected in many ways: family, society, culture, education, and economics. People are also linked to each other through their emotional lives, something that is often expressed in the arts. Art involves many aspects of creativity and requires an approach that differs from analytical thinking and uses a different logic. Another aspect of art is that while it touches on individual experience, it derives much of its power from what people have in common. It is free from scientific limitations and can give voice to our emotional life and open new avenues of thought. The different perspectives of science and art must be connected and integrated to improve our awareness of the state of the ocean and support behavioral change.

Scientists are trained in critical, analytical thinking and strive to deepen their understanding by exploring their fields in greater detail, while society has become increasingly specialized. The scientific method and engineering are strong and have made great progress in recent centuries. But society, with its management of the ocean and its coastal seas, has failed and faces increasing threats from extensive human pressures. Society expects more from the research community, and interdisciplinary programs have been initiated to address grand global challenges. Great efforts have been made to formulate and take steps toward a sustainable global path. Here, the UN has played an important role by developing a strong vision for international sustainable development. These goals require significant efforts to rethink environmental management, political ambitions, and human and ocean values.

In recent decades, natural science has identified many threats to the marine environment, as discussed in this book. Natural scientists need support from the human sciences to provide opportunities for new ways of thinking and acting. The union of natural and human sciences enables much more joyful, pluralistic, and creative thinking that can also lead to changed behavior.

The ocean's services to humanity are immense, fundamental, and invaluable. Attempts to put a monetary price on the ocean and its services are doomed to be misleading. A change in human's attitude to the ocean and the environment must be based on something other than simplified and reductive economic calculations. The

A. Omstedt, *A Philosophical View of the Ocean and Humanity*,
https://doi.org/10.1007/978-3-031-64326-2

necessary change in attitude should instead be a fundamental change in our understanding of human values and how we interact. With a growing global population and increasing urbanization, competition for resources and space competition will increase. This can lead to alienation and increased ocean destruction. Climate change scenarios warn us about a future with increased fragmentation. A green global world that can address, rather than surrender to, future challenges will require new values.

The book's first part introduced the reader to how analytical thinking and intuition can be connected. Here, we use photos, poetry and dreams to examine how fleeting feelings can be translated into stories of great value. Our imagination and our dreams form the basis of human creativity. When society has an overconfidence in rationality, we are deprived of the possibility of developing these inner resources. The first part of the book aimed to increase the understanding of intuition and how we think and give the reader a broad understanding of where not only facts and values influence us but also the imagination. The starting point is the author's deep dive into his intuition and dream world, which can also be compared to an inner ocean of gigantic possibilities. To be freed from feelings of being frozen and trapped by expectations, science, literature, and dreams are explored to find knowledge, feelings, and inspiration. In the next part of the book, what we learned in part one was applied, mainly through a better understanding of image interpretation, art, and intuition when we reflect on how we can change our behavior toward the ocean.

"*If you ask the ocean, the ocean will give you the answer*," a poem quoted at the beginning of part two of the book. Some of the ocean's responses convey great confidence in humanity, even though most of us today have lost our ability to connect and communicate, ignoring human values meaningfully. We are reminded that the most meaningful part of life is nurturing new life under healthy living conditions. Developing healthy and joyful living conditions is possible, and with the growing human population, it should be humanity's primary task. Destructive thinking and behavior scare us and will not promote the future development of a more sustainable lifestyle. Human development and culture brought about by art, language, and science have extraordinary capacities and can provide the impetus for necessary mental change, as illustrated by humanity's long history of daring achievement. Nothing, except human nature, has been too difficult to overcome, but the price has often been high. Each new generation builds on a long tradition of previous achievements, and each new generation must learn that when people are free from fear, others become braver, open-minded, and generous.

Humans need a vision to unleash their imagination and resources while keeping their feet planted on the earth. The most important vision can be to help each other and not destroy the ocean, sway freely like a kelp forest while sharing a vision and serving society in the interest of future generations. Doing so takes courage. Considering all that people get from the ocean, it is clear that people get more than they give and that feedback and partnership require more generosity. The best way to overcome traumatic feelings may be to sleep and cherish dreams. So many people need to start sleeping more consciously and realize that within all people, there is an imperishable core of being, a part that needs to be shared to improve our health. Our collective knowledge has such potential and invents increasingly environmentally friendly processes in the areas of

- Energy supply,
- freshwater production through desalination,
- food production from a healthy ocean,
- green cargo transport,
- green urbanism,
- organic farming,
- mental and physical health.

If we manage to improve our thinking by better connecting analytical thinking and intuition, it will strengthen our courage to tackle global challenges. We must find joy in learning more about our beautiful blue planet and ourselves. So much of the true significance and potential of humanity and the ocean remains unrealized. The health of the ocean and humans will depend on harmony achieved through the cooperation of many people to create something greater, as when an orchestra gives something more than the individual instruments can, as in the famous music *La Mer* by Claude Debussy from early 1900. Collaboration beyond and across separate disciplines can create a new story that brings the ocean and humanity into a richer relationship.

In part three of the book, the topic was broadened. The development of the ocean and the earth has been going on for billions of years and formed a blue planet with unique conditions for life. Only recently have humans come to influence nature in many negative ways. However, today, there is a completely different knowledge and conditions than previous generations, which means we can better manage the oceans. The ocean cannot take care of itself, and we cannot restore it to how it was under previous conditions. We must realize that the ocean's future is closely linked to what humans do. Facts are not enough here, but I highlight in this book the need to connect science and art or facts and feelings. By paying attention to the voices of the ocean embodied in art and collected through extensive observations by marine scientists, there are opportunities for a new relationship. With access to the scientific source of knowledge and inspiration from the artistic source of knowledge, there is hope and insight that a fundamentally new way of thinking is possible, where humans and the ocean can be united as guarantors of vitality and progress.

So, to you who read this book, I want to convey the courage to dive into the inner and outer oceans. There are unlimited amounts of hidden intelligence to discover and be inspired by. There are also solutions to create a new relationship with the ocean and find harmony that is greater than sustainability. All the voices of the ocean today call for our attention.

Questions

The following questions are for those who want to think further—individually or in a group. At the same time, the questions highlight central ideas in the different parts of the book and can, therefore, illustrate the book's content. Some questions are more factual, while others investigate values and feelings. Test the questions, choose those you are curious about, and feel free to work in groups.

Part I and the Night Abounds with Inspiration
Analytical and critical thinking are important aspects of our thinking that are trained in various educations and involve several important elements such as identifying the main arguments, evaluating these arguments, and identifying hidden agendas or flaws in the argumentation. Intuition is an ability that everyone has that can generate something new and unexpected and can also be trained. Analytical thinking is often a slow process, while intuition is fast. To combine analytical thinking and intuition, we need to work with both aspects but also integrate them into a relevant story. Below are some exercises to train the connection between analytical and intuitive thinking. But first, some questions on part I.

What is rational thinking, intuition, and dreams?
How and why should one connect analytical and intuitive thinking?
What is the difference between open and closed questions?
What are triggers, and how do they play into our emotions and dreams?
What do symbols and metaphors describe?
What happens when you put feelings into words?
What happens when you make a dream, a work of art, or a problem your own?
How can you bring analytical thinking and intuition together?

Exercise 1

Choose any illustration in this book. Keep it in your mind and put away the illustration. Write down a short description of the picture. *Tip*: Stay in a room where you can concentrate, sit quietly in a chair, and focus on yourself; study the picture, closing

A. Omstedt, *A Philosophical View of the Ocean and Humanity*, https://doi.org/10.1007/978-3-031-64326-2

your eyes if necessary before describing the image. Here, you train your memory and ability to observe details. You often get a better connection with your creativity by closing your eyes.

Exercise 2

Identify different parts of the image and symbols based on your memory. Explore what kind of feelings the image evokes in you. Write down your feelings and explore possible associations with the various symbols. *Tip*: An image with a burning candle can create feelings such as calm, warmth, presence, aloneness, happiness, etc. The light can symbolize guidance, wholeness, hope, and, for example, peace. Make the image your own by thinking it illustrates a part of your life and asking what feelings it creates and what the symbols and metaphors indicate. Here, you practice putting feelings into words, an important step in learning to be involved. You also practice understanding something deeper than just the object, thus practicing empathic thinking. A light is a light, but it can be understood in many different ways.

Exercise 3

Explore your life situation by writing down your feelings about these exercises before you start reading this book, yesterday, a week ago. *Tip*: You probably didn't choose the picture by chance but for a reason. In all life situations, we carry many emotions from the past that influence how we react. You practice observing your feelings and how these can influence your decisions here.

Exercise 4

Read back what you have written and add any new ideas about the image and why you chose it. *Tip*: Read slowly and reflect on what emerges from the text and your life situation. You practice deepening your analytical thinking and strengthening your intuition through your reflections here. You often find more when you read about what you wrote. Here, it illustrates the value of thinking more deeply, respecting your feelings, and giving thought the necessary time.

Exercise 5

Write down a summary based on the picture and your reflections. Put a title on the text. *Tip*: Write the summary slightly dramatically and let the title summarize your view of what was created by the image. You learn to effectively combine analytic thinking, life situations, and intuition here. This exercise trains your ability to gather all the information rationally. This is similar to what is expected from a chairperson who has listened to all the arguments when summarizing the meeting. You practice putting together what has been written down and what has not been spoken in a new and creative way.

Exercise 6

Choose a question that you want to deepen your understanding of. Sketch an illustrative figure of the problem you want to address. Examine your feelings and reasons for choosing this problem. Examine your life situation regarding skills you need to improve, the available time, and guidance from the student environment and the library. Formulate a short title for your problem, such as a question. Write down a summary, including a sketch of the problem. Consult good teachers/researchers. *Tip*: A group can be very helpful when dealing with new issues. You can ask them: If this is your problem, how could you solve it? As you explore the question, write down discoveries and feelings by expanding your summary into an article structure. Organize your thinking by, for example, preparing a PowerPoint presentation or artistic sketch. A big step toward success is making the problem your own. You will deepen your understanding by focusing on the question as your project. Be open to new ideas and feelings and push thinking to its limits.

Part II in Search of a New Ocean Relationship

This part overviews how human behavior threatens our beautiful blue planet. But also tools for connecting facts and values to change our relationship with nature. In dialogue with the ocean, we are stimulated to examine our values and strengthen qualities such as compassion, curiosity, empathy, and creativity. These properties are crucial for opening a new relationship with the ocean—first, some questions on Part II.

How and why does plastic end up in the ocean? How does it spread, and what happens with the plastic and the environment?
How is ice formed in the sea, and why does ice float?
What drives currents?
How are eddies and turbulence formed in the ocean?
Where does all the freshwater come from, and where is it stored?
Describe how evaporation in the tropics affects the large-scale winds in the atmosphere and currents in the ocean.
How does the greenhouse effect work, and what are the most important greenhouse gases?
Where does all the salt in the ocean come from?
How old are the earth and the ocean, and when were the first organisms formed?
How did the first life form on earth?
What is the difference between human and octopus intelligence?
How are phytoplankton formed, and what do they mean for the ocean and the atmosphere?
Give examples of some different marine ecosystems and what can change them.
What are the risks of overfishing for fish stocks and the ecosystem?
Give examples of non-living marine resources.
How is mineral oil formed?
Why is the term ecosystem services misleading?
Give examples of observations that illustrate that humans affect the ocean climate.
Give examples of other human threats to the ocean.
What is meant by human and ocean health?

What is the difference between a sustainable relationship and a relationship in harmony?
Can we talk about the intelligence of the ocean and what, if any, could it be?

Exercise 7

Doctor Stockmann finds that bacteria seriously contaminate the bathhouse water in Henrik Ibsen's play *An Enemy of the People.* When he conveys this to society, he comes into conflict with everyone. Could Dr. Stockmann have acted differently?

Exercise 8

Continue in the group and reflect on what is needed to change human behavior toward the ocean and its coasts. What is the role of marine scientists in collecting data? Environmental laws? Monitoring? And what is missing?

Exercise 9

Go outside and sit by a beach. Sit comfortably and look out to sea. What emotions are awakened in you? Sit quietly and reflect on the threats to which humans expose the ocean. What emotions are aroused in you? How do emotions affect your thinking?

Exercise 10

The population is increasing, and more and more people are moving to cities. Will alienation from the ocean increase? How is this countered? How can we better listen to the voices of the ocean? How can we representing the ocean's voice by measurements, models, sounds, art, literature, and dreams? How can we open a new relationship?

Exercise 11

The chapter titles balance facts and feelings such as curiosity, service, vulnerability, interpretation, belonging, partnership, courage, listening, hope, vision, harmony, and simplicity. Reflect on how these values can change our relationship with the ocean.

Part III Science, Art, and Inspiration
In this part, the description of man's relationship to the ocean and our thinking is deepened with inspiration from literature and philosophy. Various possibilities for changing behavior are discussed there.

Exercise 12

There are many dystopias, i.e., stories that paint a negative social development and the end of the earth. What emotions do they evoke, and can they lead to change? What is a better approach to dystopian sentiments?

Exercise 13

How does Aristotle think that knowledge can be divided into? When new and unknown problems appear, what kind of knowledge do we have access to, and how do we create credible measures?

Exercise 14

Choose any image, text, photograph, or illustration from the book. Examine your sources of knowledge by reflecting on what you see if you look at the image from a scientific, artistic, or mythological perspective.

Exercise 15

Why should we care about the future of the ocean?

Exercise 16

Give examples of stories and music that influenced your relationship with the ocean.

Exercise 17

Our World in Data (https://ourworldindata.org/) is a free database where you can search for knowledge about population development and global problems such as hunger and poverty, as well as about climate change and much more. Familiarize yourself with this database and examine how global population growth and energy consumption are developing. Feel free to search for anything else that interests you.

Glossary

ACIA Arctic Climate Impact Assessment

Acidification Decrease of pH in water

Accumulation time The time it takes to fill a reservoir

Agenda 2030 Adopted by the United Nations in 2015, it summarizes 17 global goals for sustainable development

Albedo The fraction of solar radiation reflected by a surface or object

Anoxic Free of oxygen

Anthropogenic Resulting from human activity

BACC BALTEX Assessment of Climate Change

Baltic Earth Network for improved earth system understanding of the Baltic Sea region

BALTEX Baltic Sea Experiment

Biochemical processes Chemical processes that occur in living organisms

Biogeochemical processes Chemical processes that occur in the living and non-living compartments of the studied system

Biosphere The sum of all the earth's ecosystems

Climate The statistical state of weather over long periods, typically at least 30 years

Climate variability Changes in climate over space and time attributable to many factors

Climate change Change in climate that is attributed to human activity

Chlorophyll Green pigments responsible for allowing plankton to absorb energy from light

Compassion The ability to notice what others need, desire, and actions to prevent or mitigate suffering

Conscious mind The part of the mind that is aware of external or internal objects

Courage The ability to do something that frightens one

Curiosity The desire to know or learn something

Cyanobacteria A special type of phytoplankton, also called blue-green algae

DDT Dichlorodiphenyltrichloroethane, a pesticide used in agriculture and households

A. Omstedt, *A Philosophical View of the Ocean and Humanity*,
https://doi.org/10.1007/978-3-031-64326-2

Desalination Removal of salt from seawater to generate freshwater
Diatoms One group of phytoplankton
Downwelling Sinking surface water in a water body
Dreams A series of thoughts, images, and emotions occurring during sleep
Ecosystem A community of organisms interacting with the non-living parts of the environment
Ekman transport Wind-driven water transport 90° to the right (left) of the wind direction in the northern (southern) hemisphere
Elastic deformation Reversible change in the shape of a material (e.g., a rubber band)
Eukaryotes Organisms whose cells have a nucleus enclosed within membranes
Eutrophication Addition of nutrients to a water body
Evolution Development over time and generations of different kinds of organisms
Evaporation The transition of liquid water to the gas phase
FAO Food and Agriculture Organization of the United Nations
Fragile material Easily broken or damaged material, such as glass or sea ice
Frazil ice Crystals of ice formed in turbulent super-cooled water
Freshwater content The amount of freshwater available in ocean water
Future Earth A program and network to explore risks posed by global environmental change and opportunities for sustainability
Geophysical fluid dynamics Fluid dynamics of naturally occurring flows on the earth and other planets
Glaciation The formation, movement, and recession of glaciers
Global mean temperature Surface air temperature averaged over the globe during a given period
Gravity Force by which a planet or other bodies draws objects toward its center
Greenhouse gases Water vapor, carbon dioxide, methane, nitrous oxide, and ozone
Harmony Collaboration between different "voices" to create something greater
Heat capacity Equal to the ratio of the heat added to the temperature change
Heat conduction The movement of heat between parts of a substance that have different temperatures
Hidden intelligence Unconscious processes that underlie action and thinking
Humanity The sum of all humans
ICES International Council for the Exploration of the Sea
Integrity The practice of being honest
Interdisciplinary research Combines research in two or more academic disciplines into a single activity, creating new approaches by thinking across boundaries
Intuition The ability to understand something instinctively
IPCC Intergovernmental Panel on Climate Change
ISA International Seabed Authority
Kelp forests Underwater areas populated with large brown algae of the kelp species
Latent heat Energy required for phase change, such as freezing, melting, evaporation, or condensation
Latent heat flux Heat flux associated with freezing, melting, evaporation, or condensation

Listening Attention to internal and external voices
Long-wave radiation Energy radiation emitted by all bodies, such as the earth and clouds
Metaphor A statement or image regarded as representative of something else
NAO North Atlantic Oscillation
Net precipitation The difference between precipitation and evaporation
Ocean Decade The UN Ocean Decade (2021–2031) program
Pack ice Ridged sea ice
Pancake ice Rounded pieces of ice formed from frazil ice and waves
Parameterization Defining or choosing parameters that describe unresolved processes
Perception The ability to see, hear, or become aware of something through the senses
pH A number expressing the proton concentration (H^+) of a solution on a logarithmic scale
Plastic deformation Non-reversible change in the shape of a material (e.g., glass or sea ice)
Prokaryotes Unicellular organisms that lack membrane-bound organelles and a defined nuclei
Regime shift An abrupt and persistent change
Removal time The time it takes to empty a reservoir
Sensible heat flux Heat flux associated with the temperature change of a body or system
Short-wave radiation Energy radiation at wavelengths in the visible, near-ultraviolet, and near-infrared spectra
Stratification, seawater Differences in density between layers of seawater
Stochastic Random
Stoichiometry Calculation of reactants and products in chemical reactions
Subconscious mind The part of the mind not currently in focal awareness
Sustainability Using resources mindfully so that their supply never runs out
Sverdrup transport Theoretical relationship between wind stress and the vertically integrated meridional transport of ocean water
Symbol A sign that is representative of something more than what it indicates
Thermohaline circulation The part of large-scale ocean circulation driven by global density gradients
Thermodynamics Field of science dealing with heat and temperature and their relationship to energy and work
Trans-disciplinary research Several disciplines working jointly and moving beyond discipline-specific approaches to address a common problem
UN United Nation
Unconscious mind Processes in the mind that occur automatically and are not directly available to introspection
UNCLOS United Nations Convention on the Law of the Sea
Upwelling Rising of deep water to the surface of a water body

Viscous media Fluids that resist deformation by shear stress (i.e., motion of layers differing in velocity)

Vision Ability to think about, imagine, or plan a future

Water balance Equation summarizing all the in- and outflows of water to and from a water body